Anniversary Feature Papers

Anniversary Feature Papers

Editor

Steven Y. Liang

MDPI • Basel • Beijing • Wuhan • Barcelona • Belgrade • Manchester • Tokyo • Cluj • Tianjin

Editor
Steven Y. Liang
Georgia Institute of Technology
USA

Editorial Office
MDPI
St. Alban-Anlage 66
4052 Basel, Switzerland

This is a reprint of articles from the Special Issue published online in the open access journal *Journal of Manufacturing and Materials Processing* (ISSN 2504-4494) (available at: https://www.mdpi. com/journal/jmmp/special_issues/anniversary_feature_papers).

For citation purposes, cite each article independently as indicated on the article page online and as indicated below:

LastName, A.A.; LastName, B.B.; LastName, C.C. Article Title. *Journal Name* **Year**, *Volume Number*, Page Range.

ISBN 978-3-0365-0188-8 (Hbk)
ISBN 978-3-0365-0189-5 (PDF)

Contents

About the Editor

Steven Y. Liang, Ph.D. in Mechanical Engineering from the University of California at Berkeley, USA. Currently, he is the Morris M. Bryan, Jr. Professor for Advanced Manufacturing Systems at the Georgia Institute of Technology. His technical interests lie in physics-based computational mechanics of predictive manufacturing processes. Prof. Liang served as President of the Walsin Lihwa Corporation (publicly traded), President of the North American Manufacturing Research Institution (NAMRI), and Chair of the Manufacturing Engineering Division of the American Society of Mechanical Engineers (MED/ASME). He is currently serving as Technical Editor of the International Journal of Precision Engineering and Manufacturing (Springer) and Editor-in-Chief of the Journal of Manufacturing and Materials Processing (MDPI). Prof. Liang has published over 700 archival scientific articles and 5 books. Among his accolades, he has received the Robert B. Douglas Outstanding Young Manufacturing Engineer Award of the Society of Manufacturing Engineers (SME), the Ralph R. Teetor Education Award of the Society of Automotive Engineers (SAE), the Blackall Machine Tool and Gage Award of the ASME, and the Milton C. Shaw Manufacturing Research Medal of the ASME. Prof. Liang is also a fellow of the ASME, SME, and Academy of Engineering and Technology (AET).

Preface to "Anniversary Feature Papers"

The *Journal of Manufacturing and Materials Processing* (JMMP) aims to provide an international forum for the documentation and dissemination of recent, original, and significant research studies in the analysis of processes, equipment, systems, and materials related to material heat treatment, solidification, deformation, addition, removal, welding, and accretion for the industrial fabrication and production of parts, components, and products. The *JMMP* was established in 2017 and has published 14 issues and more than 300 contributions. It has been listed in the ESCI, Inspec (IET), and Scopus (Elsevier).

In celebration of the anniversary of the *Journal of Manufacturing and Materials Processing*, the Editorial Office has put together this Special Issue, which includes several representative papers that reflect the vibrant growth and dynamic trend of research in this field:

(1) Establishment of advanced and innovative manufacturing methodologies—as presented in the papers entitled "Investigation on Product and Process Fingerprints for Integrated Quality Assurance in Injection Molding of Microstructured Biochips" and "Machining Forces Due to Turning of Bimetallic Objects Made of Aluminum, Titanium, Cast Iron, and Mild/Stainless Steel";

(2) Processes to transform material properties and characteristics for subsequent manufacturing steps to be performed—as discussed in the papers of "Temperature- and Time-Dependent Mechanical Behavior of Post-Treated IN625 Alloy Processed by Laser Powder Bed Fusion" and "Finite Element Modeling of Orthogonal Machining of Brittle Materials Using an Embedded Cohesive Element Mesh";

(3) Design of equipment or the development of tooling for materials processing and manufacturing—as given in "Five-Axis Machine Tool Coordinate Metrology Evaluation Using the Ball Dome Artefact Before and After Machine Calibration" and "Flexible Abrasive Tools for the Deburring and Finishing of Holes in Superalloys";

(4) Assessment and control of process quality, efficiency, and competitiveness—as explored in the papers of "Performance Comparison of Subtractive and Additive Machine Tools for Meso-Micro Machining" and "Thermal Modeling of Temperature Distribution in Metal Additive Manufacturing Considering Effects of Build Layers, Latent Heat, and Temperature-Sensitivity of Material Properties";

(5) Capability enhancement of materials processing and manufacturing through prediction, modeling, analysis, optimization, monitoring, and control—as outlined in the deliberations of "Optimization of Laser Powder Bed Fusion Processing Using a Combination of Melt Pool Modeling and Design of Experiment Approaches: Density Control", "Discrete Element Simulation of Orthogonal Machining of Soda-Lime Glass with Seed Cracks", and "Effect of Post Treatment on the Microstructure, Surface Roughness and Residual Stress Regarding the Fatigue Strength of Selectively Laser Melted AlSi10Mg Structures".

This Special Issue shows that manufacturing and materials processing is an actively growing area in the research community. The scope and the findings of work presented in the *JMMP* have carried both palpable scientific merits and tangible application relevance. As the needs in this area continue to rise, it is expected that the interest in research will expand and the outcomes from studies will flourish in the future.

The success of the *JMMP* is attributed to all the scientific authors for their outstanding contributions. Sincere appreciation is due to the peer reviewers for their constructive comments and suggestions and also to the editorial team for their commitment in facilitating the high-efficiency and high-quality operation of the journal.

Steven Y. Liang
Editor

Journal of
Manufacturing and
Materials Processing

Article

Investigation on Product and Process Fingerprints for Integrated Quality Assurance in Injection Molding of Microstructured Biochips

Nikolaos Giannekas *, Yang Zhang and Guido Tosello

Department of Mechanical Engineering, Technical University of Denmark, Produktionstorvet, Building 427A, DK-2800 Kgs. Lyngby, Denmark; yazh@mek.dtu.dk (Y.Z.); guto@mek.dtu.dk (G.T.)
* Correspondence: nikgia@mek.dtu.dk; Tel.: +45-4525-4747

Received: 8 October 2018; Accepted: 12 November 2018; Published: 15 November 2018

Abstract: Injection molding has been increasing for decades its share in the production of polymer components, in comparison to other manufacturing processes, as it can assure a cost-efficient production while maintaining short cycle times. In any production line, the stability of the process and the quality of the produced components is ensured by frequently performed metrological controls, which require a significant amount of effort and resources. To avoid the expensive effect of an out of tolerance production, an alternative method to intensive metrology efforts to process stability and part quality monitoring is presented in this article. The proposed method is based on the extraction of process and product fingerprints from the process regulating signals and the replication quality of dedicated features positioned on the injection molded component, respectively. The features used for this purpose are placed on the runner of the moldings and are similar or equal to those actually in the part, in order to assess the quality of the produced plastic parts. For the purpose of studying the method's viability, a study case based on the production of polymer microfluidic systems for bio-analytics medical applications was selected. A statistically designed experiment was utilized in order to assess the sensitivity of the polymer biochip's micro features (μ-pillars) replication fidelity with respect to the experimental treatments. The main effects of the process parameters revealed that the effects of process variation were dependent on the position of the μ-pillars. Results showed that a number of process fingerprints follow the same trends as the replication fidelity of the on-part μ-pillars. Instead, only one of the two on-runner μ-pillar position measurands can effectively serve as product fingerprints. Thus, the method can be the foundation for the development of a fast part quality monitoring system with the potential to decrease the use of off-line, time-consuming detailed metrology for part and tool approval, provided that the fingerprints are specifically designed and selected.

Keywords: precision injection molding; quality control; process monitoring; process fingerprint; product fingerprint

---

1. Introduction

In the last decades, the development of new technology, legislation, and customer needs have influenced a change in the functional requirements and design of complex parts, while keeping the focus on high volume mass production processes that maintain a cost-efficient production for many applications. Such applications originate in the automotive, electronics, communication, and medical industries, as well as in micro manufacturing [1,2]. A process that can maintain a cost-effective production with short cycle times is injection molding. Injection molding is continuously gaining market share in the production of cost effective products, accounting for 50% of the produced plastic parts [3], in comparison to other manufacturing processes.

In a plethora of industrial sectors, and particularly in the medical sector where biomedical and drug delivery devices are concerned, applications with integrated μ-features, such as μ-pumps and μ-measuring devices for the precise handling and administration of drugs, dictate the need for tight tolerances in order to satisfy the functional requirements of the product [4]. Such functional requirements are challenging to fulfil for all the injection-molded components in a high-volume production. They require a stable process with frequent metrological inspections in order to ensure process stability and high part quality. Metrological studies though require a significant amount of time in comparison to the cycle time of injection molding, which is often in the order of few seconds. Due to the high costs involved, especially in the cases of micro molding equipment and micro tools for μ-applications or applications with μ-features, process monitoring is an attractive research subject. The main objective is the monitoring of the process for the occurrence of defects and quality assurance of the molded parts, since an out of tolerance production can lead to an inefficient production line with high costs and scrap rate.

The current paper presents an alternative approach to continuous or statistical monitoring and part quality control, by proposing indexes that serve as part quality indicators (QI) (i.e., "product and process fingerprint") based both on process and product data.

The presented approach is developed in two parallel tracks. Firstly, the "product fingerprint" track which considers the use of dedicated μ-features positioned on the runner of the component that are equal or similar in size and shape to the features on the part [5]. The two sides of the microfluidic system are used as a study case. The μ-pillars positioned on the microfluidic system are designed as functional micro features [6] that direct the flow of the liquid and inhibit the formation of air bubbles. As functional features, their replication fidelity is of high importance for the overall quality and acceptance of the microfluidic component. The correlation of the features' replication on the runner to the ones in the part is going to be explored. Current research presents numerous examples of part features in use for fast part quality inspection. Two prominent examples are the use of weld line position to assess the quality of the molded part as described by Tosello et al. [7], and the use of nano-features placed on different areas of a component that provide the necessary indicators for fast part quality assessment as discussed by Calaon et al. [8]. However, in both those cases the μ-features are positioned in the cavity.

The "process fingerprint" track investigates the suitability of the transient time-resolved process data originating from the injection molding machine control sensors, for process monitoring and consequently part quality control. A number of researchers in the field of sensor technology have studied different approaches to develop methods of process control, an optimization that could shorten the duration of metrological investigations for the approval of injection-molded components. Promising results are shown in studies where in-mold sensors are used for process regulation and monitoring, though the placement of sensors involves higher tooling costs [9–13]. Chen et al. [14] have proved that part weight and thickness can be reliably monitored with the use of a linear variable differential transformer (LVDT) monitoring the mold separation (MS) distance. Instead Gao et al. [12] have developed a custom multivariate sensor (MVS) in order to monitor the quality on the injection-molded parts based on the hypothesis that part quality indicators (dimensions) can be tightly controlled and the in-mold process parameters are already known.

Further studies are using data from external sensors placed on the mold or in-line measuring equipment to monitor and optimize the process considering the component's functional requirements. An online multivariate optimization system for the optimization and control of the process has been developed by Johnston et al. [15], while Yang et al. [16] have detected defects in the process with the use of an in-line digital image processing method. Consequently, for the detection of a defect, the software feeds data to a process optimization algorithm built on a model-free optimization (MFO) procedure. Other approaches involve the use of numerical simulation procedures for the monitoring and optimization of the process, such as the work on dynamic injection molding and sequential optimization of warpage based on the Kriging surrogate model, presented by Wang et al. [3], and the application of artificial neural networks (ANNs) and genetic algorithms as discussed by Ozcelik et al. [17].

Most of the approaches discussed in literature focus on tightly controlled and optimized processes, with the dimensional control of the injection-molded components to be indirectly considered. However, the main target of any quality control system is the quality of the final product, and thus coupling the replication fidelity of the parts to the sensor data is a requirement.

The current paper presents an alternative approach based on process and product fingerprints. The remainder of the article is structured as follows: in Section 2 the experimental setup and methods are presented; in Section 3 the results are discussed; in Section 4 a summary of the article and conclusive remarks are given. The extraction of both process and product fingerprints is discussed with the selection of the most suitable "fingerprints" to be completed.

2. Experimental Setup and Methods

2.1. Molding Tool Geometry

The experimental setup was designed in a way that accommodates both research tracks related to the process and product fingerprints. To proceed with the approach of product fingerprint and in order to access the quality of on-part micro features in correlation with on-runner μ-pillar features, specifically developed tool inserts for the production of a biochip were manufactured. The mold used was a two-cavity mold as seen in Figure 1 and the manufactured geometry consisted of the two sides of a bio-fluidic microchip for drug testing. The biochip had the form of a 20 × 20 × 2 mm plate with on-part conical μ-pillar features with 600 μm nominal height, Ø250 μm base diameter, and Ø200 μm top diameter [6] as seen in Figure 2. The tool inserts were manufactured to accommodate pillar μ-features on the runner equal to those on the part, as it can be seen in Figure 3.

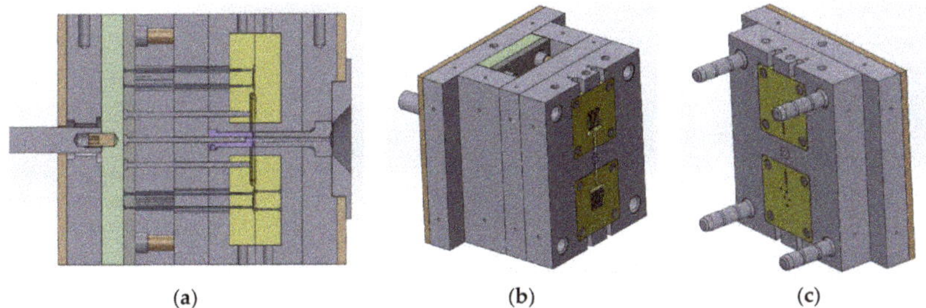

(a)   (b)   (c)

Figure 1. Half section view (**a**) and $\frac{3}{4}$ views of the movable (**b**) and stationary (**c**) sides of the mold used for the experiment.

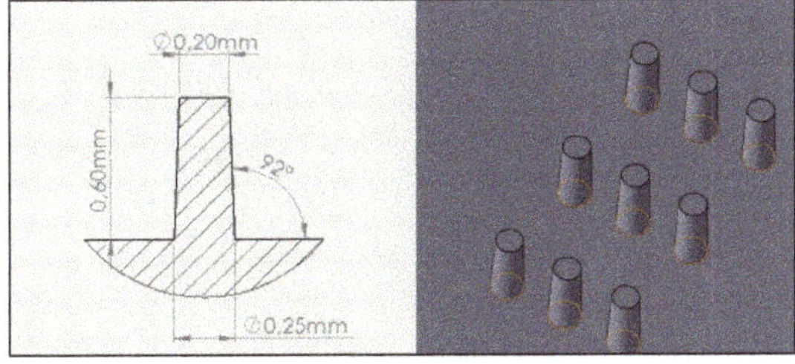

Figure 2. The micro pillars' feature shape and the dimensions of the parts.

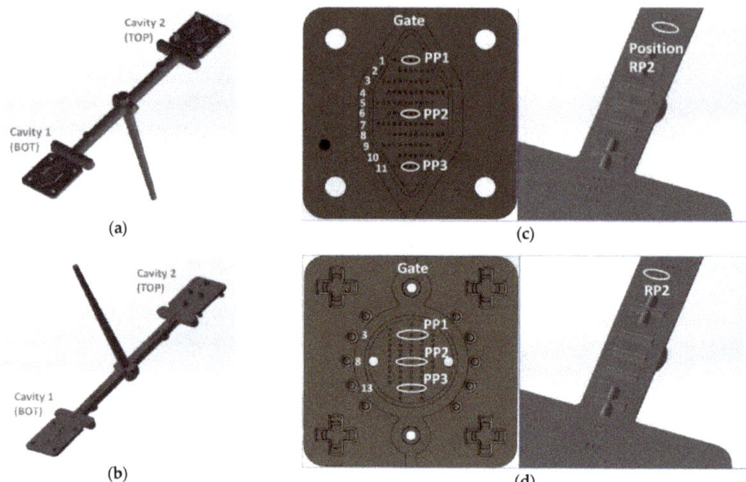

Figure 3. Molded geometry with fingerprint structures on the part and runners (**a**,**b**), and measurement positions on Cavity 1 (**c**) and Cavity 2 (**d**).

Figure 3 illustrates the geometry of the molded plastic parts and presents the positions of interest; PP1 close to the gate, PP2 in the middle of the parts, PP3 far from the gate and RP2 on the runner of the molding for both cavities. The pillars in the illustrated positions are used to assess the replication quality of the molded components for all treatment combinations in the experiments as presented in the following section of the paper. Figure 4 presents an example of the physical molded components.

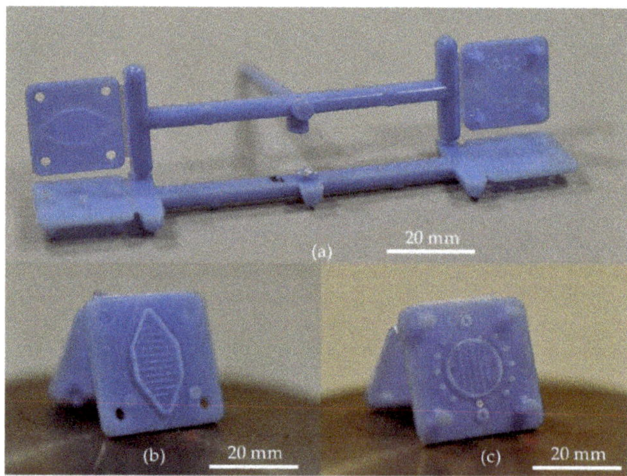

Figure 4. (**a**) Molded component with fingerprint structures on the part and runners. The fingerprints at the front (top) and back (bottom) side of the components are visible. (**b**) Bottom (Cavity 1) and (**c**) top (Cavity 2) parts of the microfluidic system.

2.2. Injection Molding Process and Experimental Conditions

The proposed product and process fingerprint concept is built on the hypothesis that the quality of the on part μ-features is correlated to the on runner μ-features and other quality indicators originated from process signals as is discussed in Sections 2.4 and 2.5. The concept requires an experimental validation to confirm the hypothesis of the micro features and extracted indices suitable to be used as quality indicators. The experiments were performed on an electric Arburg 370A injection-molding machine (Arburg GmbH + Co KG, Lossburg, Germany), with a hydraulically actuated clamping unit capable of a maximum clamping force of 600 kN and a screw whose diameter was Ø18 mm. A statistically designed $2^4 \times 3$ full factorial experiment was utilized in order to investigate the experimental process window. The parameters under consideration are: Tmelt (Tm) [°C], Tmould (Tmld) [°C], Injection Speed (InjSp) [mm/s] and Packing Pressure (PackPr) [bar] that, as from well-established research [18] and preliminary screening experiments are known to be the most significant parameters affecting the quality of injection molded components and surface replication. Table 1 presents the experimental treatments. The process parameter levels were selected by assessing the specification of the material (Figure 5), a commercial grade of acrylonitrile butadiene styrene (ABS, Styrolution Terluran GP-35, INEOS Styrolution GmbH, Frankfurt am Main, Germany), which is characterized by a relatively large processing window. Other parameters such as packing (t_{pack} = 10 s) and cooling times (t_{cool} = t_{pack} + 10 s) were set on levels high enough to avoid their influence on the responses of the experiment.

Table 1. Experimental Parameters.

Parameter	Unit	Run															
		1	2	3	4	5	6	7	8	9	10	11	12	13	14	15	16
Tm	[°C]	220	260	220	260	220	260	220	260	220	260	220	260	220	260	220	260
Tmld	[°C]	40	40	60	60	40	40	60	60	40	40	60	60	40	40	60	60
InjSp	[mm/s]	100	100	100	100	140	140	140	140	100	100	100	100	140	140	140	140
PackPr	[bar]	440	440	440	440	440	440	440	440	540	540	540	540	540	540	540	540

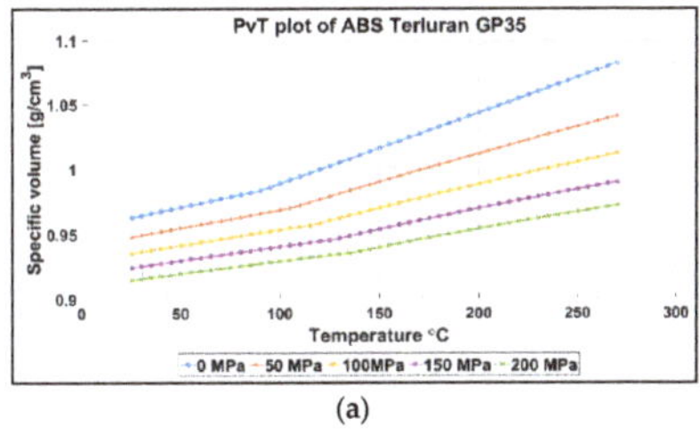
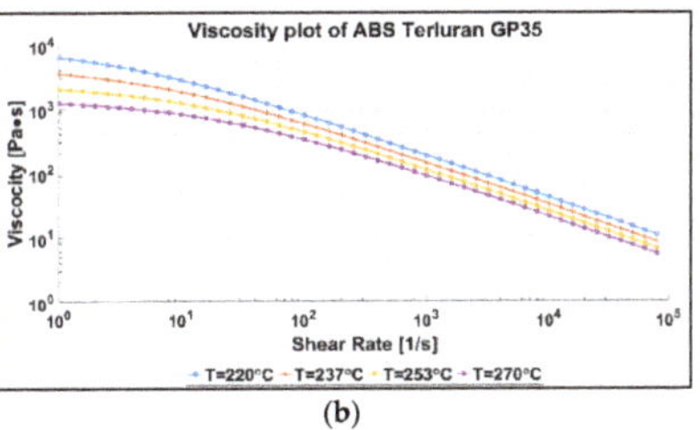

(a) (b)

Figure 5. (**a**) PvT and (**b**) viscosity plots of material Styrolution Terluran GP-35 (Acrylonitrile Butadiene Styrene—ABS) [19].

For every experimental treatment, the initial 20-molded parts from the start of the process were discarded, as the process was running to reach stability. Then the following 10 parts were collected for assessment and the three sample parts were measured (denoted as: part 1, part 5, part 10) for the assessment of the µ-pillars' replication quality and then placed both on the parts and on the runners. The sequence followed and the experiment is illustrated in Figure 6.

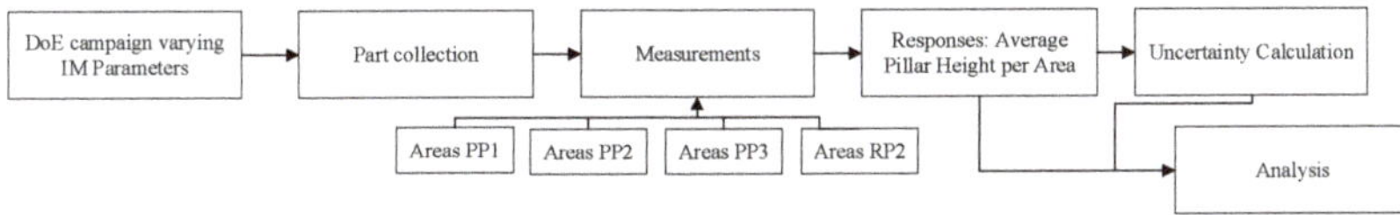

Figure 6. Flow diagram of the experimental sequence. The figure denotes the measurement areas on the part (i.e., PP1 = Part Position 1) and on the runner (i.e., RP2 = Runner Position 2) without the indication of cavity as seen in the text (i.e., Cavity 2 RP2 = C2RP2)

2.3. Pillar Dimensional Measurement and Uncertainty Evaluation Procedure

The pillar height dimensional measurements were carried out by using a focus variation microscope (Alicona Infinite Focus from Alicona Imaging GmbH, Raaba, Austria). The focus variation method is suitable for the scanning of the 3D topologies as it can effectively acquire scans of features with high slopes. A full scan of the μ-pillars though, proved to be challenging due to the almost vertical slopes (88°) of the μ-pillars. The settings used for the measurements are presented in Table 2.

Table 2. Alicona measurement settings for μ-pillars.

Measurement Settings	
Objective	×20
Exposure	3.05 ms
Contrast	1.11
Vertical resolution	299 nm

For the assessment of the process' stability, the effect of process parameter changes and the replication fidelity of the pillar μ-features for each experimental treatment, three pillars in each position were scanned to measure the μ-pillar height. The middle pillars in positions PP2 and RP2 of both cavities were measured five times in order to determine the repeatability of the measurements (standard deviation in the range of 0.1–0.2 μm was achieved) and provide sufficient data for measurement uncertainty calculations (see Section 3.1). The measurement data sets were consequently processed with the use of scanning probe image processing software (SPIP V6.4.1 by Image Metrology A/S, Hørsholm, Denmark) to extract the μ-pillar height from each scan. In SPIP, a procedure was developed to process the scans and prepare the files for pillar height calculations following the same steps for all four positions of interest by correcting the 1st order tilt in the scan as well as to set the zero background for all data-points as illustrated in Figure 7.

The average pillar height was calculated with the use of four profiles that intersected the center of the pillars with the procedure utilized to scan of both mold and molded parts in order to calculate the height and height deviation (mold-part) as a measure of the molded features replication fidelity.

To verify the quality of measurements and procedures an uncertainty evaluation was conducted. The evaluated expanded uncertainty U is a parameter associated with the measurement results and describes the data dispersion always in connection to the respective measurand. The estimation of the uncertainty and its inclusion in the replication fidelity assessment of the micro features is of great importance as the measurement repeatability and instrument accuracy can be of similar magnitude. The uncertainty budget of the measurements of the pillar heights on the parts and the respective cavity features on the mold insert were estimated based on the ISO 15530-3 (Equations (1)–(4)) [20]. The method was developed for measurements conducted with a tactile coordinate measuring machine (CMM); however, it can be adapted and applied for optical measurements [21] using Equation (4).

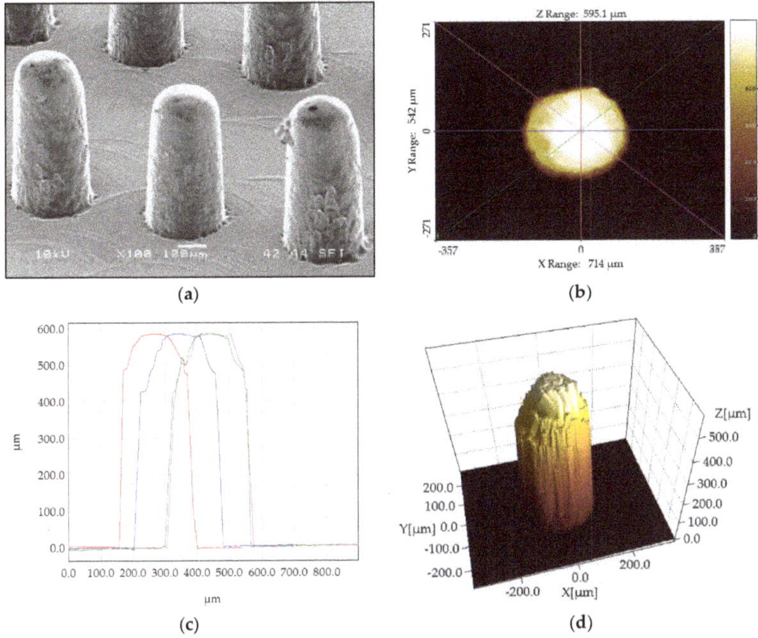

Figure 7. (**a**) SEM 3D image of the pillars and (**b–d**) pillar height measurement procedure, (**b**) step 1: extracting cros-section profiles, (**c**) step 2: assessing pillar height from the four extracted profiles as indicated by different color, and (**d**) 3D representation of the pillar [5].

The expanded uncertainty was calculated with a coverage factor k = 2 to achieve a confidence level of 95%, and four uncertainty contributors were considered (Table 3) (see Equations (1)–(3)). Such uncertainty contributors are u_{cal} which is the standard uncertainty as evaluated from a calibrated step height artefact to have traceable measurements, u_b which is the standard uncertainty associated with the systematic error (b) of the measurement process, which is the measuring instrument bias. Thirdly, the u_{th} is the standard uncertainty associated with the systematic error of the measurement process based on the heat expansion coefficient deviations of the material, since the measurements were not conducted at the reference temperature, and lastly u_p is the uncertainty associated with the manufacturing variation from either mold or parts (u_{pmould} and u_{ppart}), which is calculated using a square distribution in the modified ISO 15530-3 (Equation (4)). The measurement on individual pillars, features, and different molded parts are all affected by instrument repeatability. Thus, for u_{ppart} the maximum value of uncertainty contributor related to instrument and process is considered in order to avoid underestimating the uncertainty. These contributors are part of u_{ppart}, where: $u_{ppillar}$ is the standard deviation of five repeated measurements on the same pillar; $u_{pfeatures}$, the standard deviation of repeated measurements on four different pillar areas to estimate feature repeatability in terms of polymer replication and $u_{psample}$ the standard deviation of repeated measurements on 3 different samples on four different pillar area. The uncertainty contributors are used to calculate the uncertainty of the mold (Equation (1)) and part pillar (Equation (2)) measurements, as well as the deviation uncertainty (Equation (3)). The values of the specific uncertainties per position and experimental runs are provided in Tables 4 and 5, respectively. Table 5 provides information on the expanded uncertainty for pillar height and height deviation measurements per run.

$$U_{part} = k \times \sqrt{u_{cal}^2 + u_b^2 + u_{th}^2 + u_{ppart}^2} \tag{1}$$

$$U_{mould} = k \times \sqrt{u_{cal}^2 + u_b^2 + u_{th}^2 + u_{pmould}^2} \qquad (2)$$

$$U_{dev} = \sqrt{U_{mould}^2 + U_{part}^2} \qquad (3)$$

$$\mid u_{P_i} = \frac{data_{max} - data_{min}}{2\sqrt{3}}, \ i = \text{pillar, feature, sample for part or mold} \qquad (4)$$

$$\mid u_{ppart} = \max(u_{ppillar}, u_{pfeature}, u_{psample}). \qquad (5)$$

Table 3. Uncertainty contribution for pillar height measurements.

Uncertainty Contributions	Mold Inserts		Parts	
	Cavity 1	Cavity 2	Cavity 1	Cavity 2
u_{cal} [μm]	0.1	0.1	0.1	0.1
u_{th} [μm]	0.003	0.003	0.003	0.003
u_b [μm]	0.034	0.034	0.034	0.034
u_{ppart} [μm]	-	-	0.26–0.97	0.22–0.95
u_{mold} [μm]	0.11–0.12	0.13–0.79	-	-

Table 4. Expanded uncertainty for single pillar height and height deviations measurements.

Expanded Uncertainties	Mold Inserts		Parts	
	Cavity 1	Cavity 2	Cavity 1	Cavity 2
U_{part} [μm]	-	-	0.54–1.94	0.45–1.91
U_{mold} [μm]	0.25–0.26	0.29–1.58	-	-
U_{dev} [μm]		0.54–3.92		

Table 5. Expanded uncertainty for pillar height and height deviations measurements per Run.

	R1	R2	R3	R4	R5	R6	R7	R8	R9	R10	R11	R12	R13	R14	R15	R16
Expanded Uncertainties per Run (Uexp [μm])																
Cavity 1																
U_{part}	1.04	1.10	0.77	0.88	1.42	0.84	1.94	1.87	0.55	1.40	0.74	1.43	1.05	1.29	0.76	1.02
U_{dev}	1.65	1.74	1.77	1.86	1.86	1.77	2.09	1.69	1.98	1.95	2.02	1.83	2.49	1.68	1.80	2.28
Cavity 2																
U_{part}	1.04	1.10	0.79	0.97	1.42	0.84	1.94	1.87	1.18	1.40	1.24	1.43	1.91	1.29	0.85	1.64
U_{dev}	1.65	1.74	1.77	1.86	1.86	1.77	2.09	1.69	1.98	1.95	2.02	1.83	2.49	1.68	1.80	2.28

2.4. Product Fingerprint as Quality Indicator

The concept uses the microfluidic system described in Section 1 as a case study. It is of particular interest as it builds upon past studies that used nano features (fingerprint) on the part, where a close correlation of the fingerprint on the part to the overall quality of the component was revealed [8].

The current paper considers the use of dedicated μ-features positioned on the runner of the molding that are equal or similar in size and shape to the features on the part [5]. The μ-pillars on the runner can be used as a product fingerprint as they can be quickly measured with an in-line process set up, separated from the main component and directly correlated to the overall part quality.

2.5. Process Fingerprint as Quality Indicators

Similar to product fingerprint a set of indices is proposed to serve as QIs in order to represent the quality of the molding components with data from machine signals. Two type of QIs were considered: the first type was calculated based on the deviation of consequent signals and the second was calculated as single values per signal.

The individual QIs belonging to the first type are presented in the following sections. They were error of alignment, integrated squared error, cross correlation, shift error, and dynamic time warping. The same quality indicators were also computed for the cross-correlated signals.

2.5.1. Work of Error and Integrated Squared Error

The controller of the injection-molding machine records the injection speed and the pressure time resolved transient data during the process for every consecutive cycle. In theory, the controller and the responses of an optimized process should be the same; however, in real world conditions, the machine's controller, the components of the machine, and the material can have different behavior. For example, all operations include a level of uncertainty and interference from external conditions. As such, the recorded signals in every consequent cycle of the process can deviate from the reference cycle. This deviation describes the alignment error from each consequent cycle signal to the reference signal by Equation (6).

$$\varepsilon(t) = y_0(t) - y_i(t) \tag{6}$$

$$Er_{work} = \int \varepsilon(t)dt \tag{7}$$

$$ISE = \int \varepsilon(t)^2 dt \tag{8}$$

where: $\varepsilon(t)$ the alignment error for time instance t, y_0 the reference signal, y_i any cycle signal, and i = 1, 2, ... , N cycles.

The alignment error, though computed by Equation (6), is still a time series that consists of the amplitude difference of two signals for the time instances t, where t = 0, 1, 2, ... , 11 s. However, although the $\varepsilon(t)$ time resolved data contain valuable information, it is challenging to use. As such, the work of error (Er_{work}) and the integrated squared error (ISE) [22] as described in Equations (7) and (8), respectively, are used to extract the information as one single value for every signal associated with the deviation of each processing cycle with respect to the reference cycle. The performance of the alignment error and the ISE as a quality indicator in consequence will be discussed in a following section of the paper.

2.5.2. Shift Error

Another quality indicator is the "Shift error" or "Shift" that originates from the cross correlation of the input signals to the reference signal in every DOE run in the conducted experiment. Cross correlation in discrete time series/signals $y_0(t)$ and $y_i(t)$ is described by Equation (9) [23].

$$Shift_{y_0,y_i}(l) = \sum_{t=-\infty}^{\infty} y_0(t)y_i(t-l) \tag{9}$$

where l is the lag of signal y_i (i = 1, 2, ... , N) in association to the reference signal y_0. Cross-correlation measures the similarity between a reference y_0 and shifted (lagged) copies of y as a function of the lag as illustrated in Figure 8. The "Shift" error can be used as a QI and will be discussed further in Section 3.3.1. An example of cross correlation alignment from experimental Run 1 is provided in Figure 9.

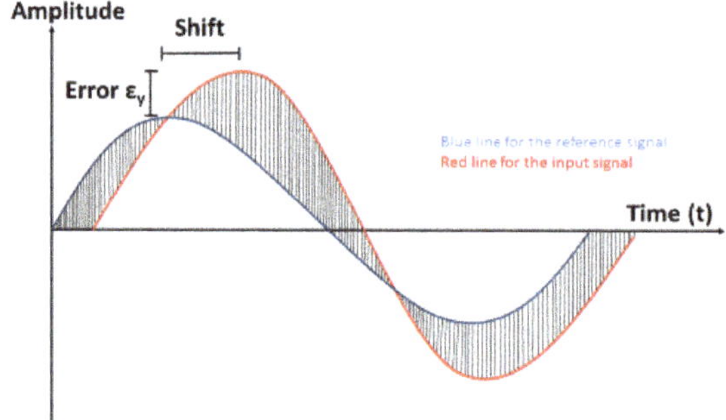

Figure 8. Alignment error ε(t) and Shift error.

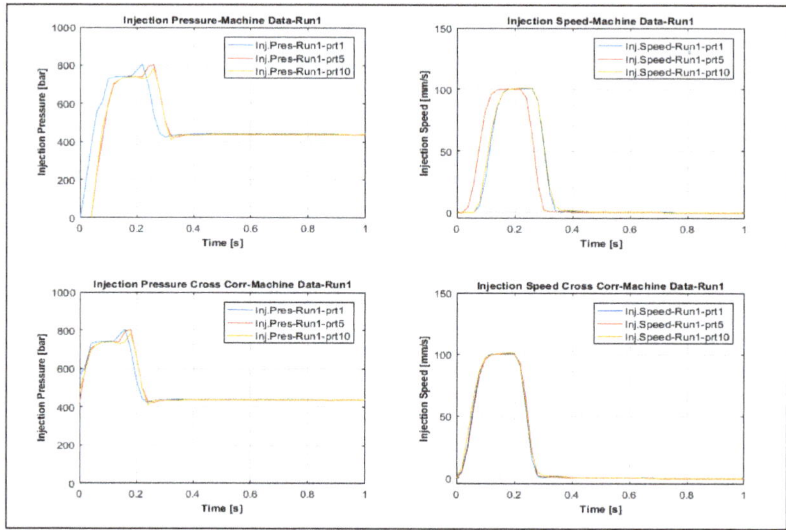

Figure 9. Signals of injection speed and pressure from part cycles 1, 5, and 10 of Run-1 before (**top**) and after cross correlation alignment (**bottom**).

2.5.3. Work Deviation

The work deviation of any consequent signal to the reference one as described in Equation (10), is an alternative QI that is used to describe similarity of any signal to the reference. A graphical representation of "WorkDev" is provided in Figure 10.

$$\text{WorkDev} = W_0 - W_i = \int y_0(t)dt - \int y_i(t)dt \tag{10}$$

where: $i = 1, 2, 3, \dots, N$ cycles.

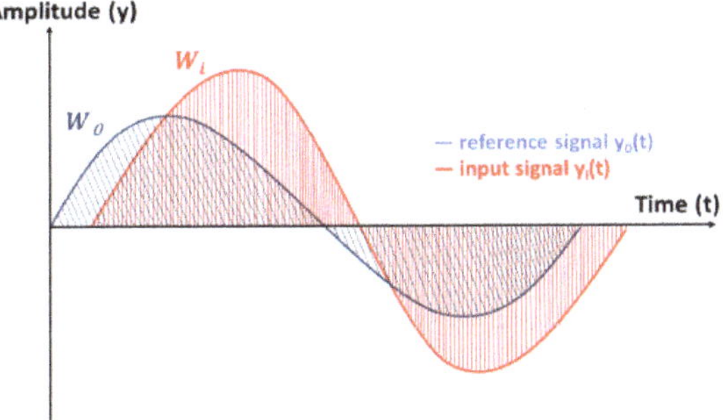

Figure 10. Representation of work deviation, given by the non-intersecting area of signals $y_0(t)$ and $y_i(t)$.

The compatibility of the "WorkDev" QI will be discussed in Section 3.3.1 and compared with the previously introduced QIs and the dynamic time warping (see next section).

2.5.4. Dynamic Time Warping

Dynamic time warping (DTW) is an algorithm that has found use in applications such as acoustics and seismic motion fields, where the alignment of a pair of time series or sequences is required [24]. The algorithm considers time series data of unequal size and it is used to compute the warping distance between two different time series or signals. The warping distance of vectors y_i to the reference vector y_0 is defined as the minimum distance from the beginning of the DTW table to the current position (k, j). Based on the dynamic programming (DP) algorithm [25] the DTW table can be calculated as follows [26] in Equation (11):

$$\text{WarpDis}: \ D(k,j) = d(k,j) + \min \begin{cases} D(k-1,j) \\ D(k,j-1) \\ D(k-1,j-1) \end{cases} \tag{11}$$

where $D(i, j)$ is the node cost connected to points $y_i(k)$ and $y_0(j)$ of the input and reference signals y_0 and y_i and is calculated with the use of L2-norm in Equation (12).

$$d(k,j) = \sqrt{\left(y_i(k) - y_0(j)\right)^2} \tag{12}$$

The warping distance ("WarpDis") is the minimum Euclidean distance in the warping DTW table.

For the purposes of this work, the single dimension DTW algorithm was used to align each consecutive signal to a reference signal. The algorithm stretches the two vectors y_0 and y_i, onto a common set of instances such that the warping distance "WarpDis", the sum of Euclidean distances between corresponding points $y_i(k)$ and $y_0(j)$, is minimized. To properly match the input and reference signals, the algorithm repeats each element of vectors y_i and y_0 as many times as necessary resulting in two signals y_i^* and y_0^* of equal size, as illustrated in Figure 11. As such, the warping distance "WarpDis" can be used as a QI.

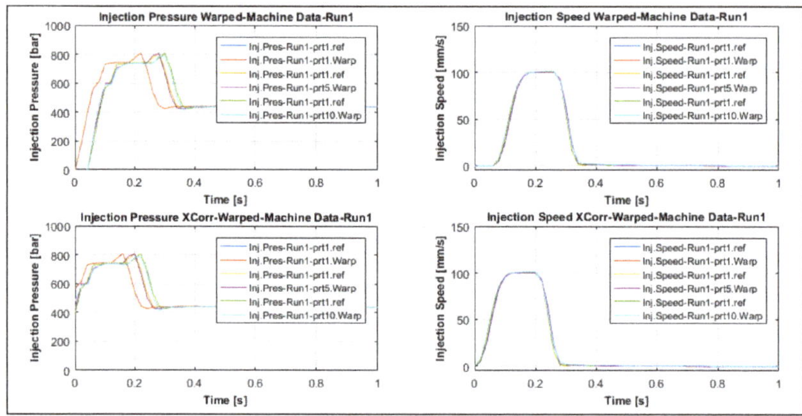

Figure 11. Alignment of original (**top**) and cross-correlated (**bottom**) signals of injection speed and pressure of Run 1 and part c ycles 1, 5, and 10 using dynamic time warping (DTW).

To ensure the validity of the previously introduced QIs, the QI values were not directly comparable to the dimensional measurements of the micro-feature on the collected samples, and the data were standardized using Equation (12).

$$\text{Zscore} = \frac{x - \mu}{\sigma} \tag{13}$$

where, "x" is the xth observation, "μ" the mean value of all observations, and "σ" the standard deviation of all observations per treatment.

Apart from the "process fingerprint" candidates originated from the deviation of both the transient injection pressure and injection speed signals to the respective reference signals, two more "process fingerprint" candidates were calculated from each signal. Those candidates belong to the second type of quality indicators and were the signal integrals and signal powers as described below.

2.5.5. Signal Integral

The signal integral "I_x" is calculated with Equation (14) and of the time resolved data from the whole signal y(t) recorded starting at the injection phase (t_0 = 0 s), till the end of the packing phase (t_n = 11 s). The integral is related to the energy stored in the system and can differ on the measured quantity. When the integral is calculated from the pressure signals, it provides the approximate value of energy stored in the polymer from the melting, compression, and injection of the molten polymer in the cavity.

$$I_x = \int_0^T y(t)dt \tag{14}$$

T: end time of signal during (time = 11 s).

2.5.6. Signal Power

The power of a signal x, "SP_x" is given as the sum of the absolute squares of the time-domain samples of the signals divided by the signal length. Similar to the integral, signal power relates to the energy of the system for all the recorded frequencies of the signal.

$$SP_x = \lim_{T \to \infty} \frac{1}{T} \int_0^T |y(t)|^2 dt \tag{15}$$

T: end time of signal during (time = 11 s).

3. Results and Discussion

3.1. Dimensional Measurements and Uncertainty Calculation

As stated in Section 2.3, three collected parts for each experimental run were examined. In order to assess the quality of the parts and of the three pillars per measurement position, as illustrated in Figure 3, they were examined to provide data for the replication fidelity of the pillars in each area of the parts and the stability of the process.

In a preliminary analysis the average pillar height (part) and pillar height deviation per area is presented in Figures 12 and 13, respectively, with their respective part measurement uncertainties as described in Equation (1) (U_{part}) and Equation (3) (U_{dev}). The uncertainty bars as illustrated on the bar graphs are associated with the combined measurement uncertainty (U_{dev}) (Figure 12) from both mold (U_{mold}) and parts (U_{part}) (Figure 10) measurements (Equation (3)) as calculated based in the ISO 15530-3 [20].

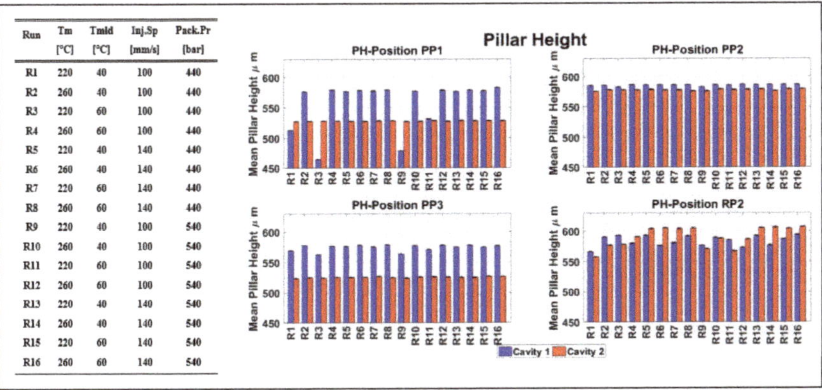

Figure 12. Average pillar height and U_{part} measurement uncertainty per position on the part for Cavities 1 and 2 parts. The *x*-axis here represents the experimental DOE runs (R1 for Run 1) as presented in Table 1.

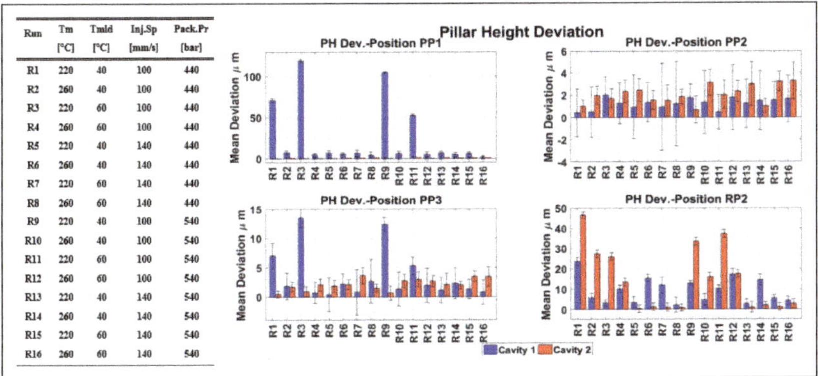

Figure 13. Comparison of average pillar height deviation (mold-part) per position for Cavities 1 and 2.

Figure 12 presents the real area pillar height of the biochip, which is homogeneous for most of the experimental runs. Figure 12 illustrates the replication fidelity of the pillars in both cavities. It is evident that the less replicated pillars are originated to position PP1, and for the experimental runs 1, 3, and 9, where the treatment uses the low value of the Tm parameter, and for runs 3 and 9 where Tmld is also at a low level. In position RP2 though, the μ-pillars positioned at the runners before Cavities 1 and 2 are better replicated for all the experimental runs where the high level of the InjSp parameter was used, as higher injection speed increased the temperature of the molten polymer through the mechanism of shear thinning. In comparison, when the lower level of InjSp was used, the replication fidelity of the μ-pillars in position RP2 was lower due to the thicker cross-section where the shrinkage was larger than the rest of the molded component.

3.2. Product Fingerprint Analysis

The dedicated μ-pillar features positioned on the runner of the molding can be potentially used as product fingerprints, as they can be rapidly measured with an in-line process set up, while already separated from the main component. However, for the on runner μ-pillars to be considered a suitable candidate for product fingerprints, sensitivity and correlation analyses are required in order to assess the sensitivity of the candidates to the process variation and their correlation to the on-part pillars, respectively.

Figure 14 presents the results of the sensitivity analysis for the effects of the process parameter changes. In particular, Figure 14a presents the results from the μ-pillar arrays height measurement in position PP1 (μ-pillar structures near the gate). From the effect plots it can be seen that the parameter with the greatest influence on the response is the injection speed (InjSp); its increase leads to 39.9 ± 3.2 μm height deviation decrease of the feature height for Cavity 1 and a 0.06 μm height deviation increase for features in Cavity 2. The error bars at the two parameter levels do not overlap, and thus, the effect is considered significant for Cavity 1. The parameter with the second most significant effect is Tm where an increase to its level results to 39.8 ± 3.2 μm height deviation (from mold values) decrease of the μ-pillars. The rest of the parameters all appear to have an influence with the exception of Tmld. However, the error bars at the parameter levels of the Tmld and PackPr parameter effects do overlap indicating that the parameters cannot be considered as significant.

Figure 14b presents the results from the pillar array height deviation measurements in position PP2 (μ-pillar structures in the middle of the part). The main effect plots reveal that the parameter with the greatest influence on the response is the InjSp, where its increase from 100 mm/s to 140 mm/s leads to 24.9 μm increase of the feature height deviation for Cavity 2, which is considered significant. For the rest of the parameters only Tm appears to have an influence; however, none can be seen as significant as the error bar in the main effect plot overlap for the two parameter levels for both cavities.

Figure 14c presents the results from the pillar arrays height deviation measurement in position PP3 (μ-pillar structures far from the gate). However, none of the effects can be considered significant as the error bars do overlap again.

In all three cases, the presented results are supported by the Pareto graphs at the right column of the figure with respect to the parameters (Tm and InjSp) that have the largest effect. The effect of the two-way interaction between Tm and InjSp is smaller from the effects of those two parameters, thus it is considered insignificant.

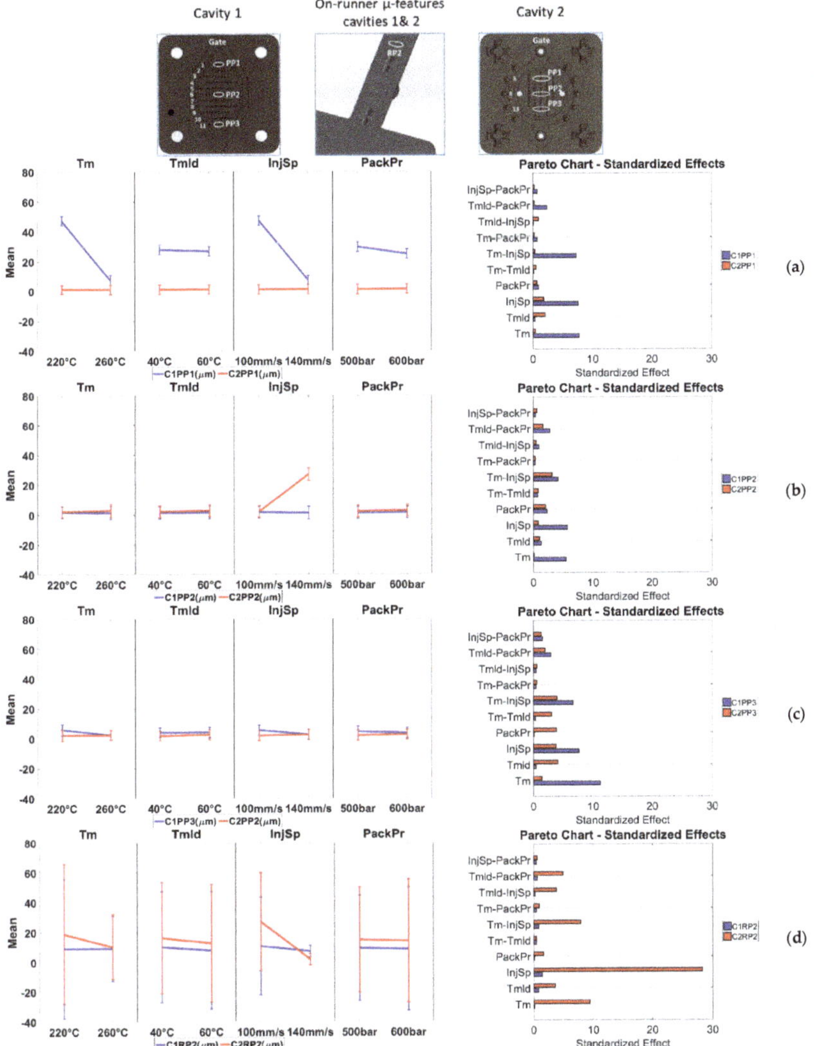

Figure 14. Influence of IM process on the eight measurand deviations (from mold values) and "product fingerprints candidates". (**a**) Position PP1 in Cavities 1 and 2, (**b**) position PP2 in Cavities 1 and 2, (**c**) position PP3 in Cavities 1 and 2, (**d**) position RP2 in Cavities 1 and 2. The figure presents the main effects (left column) and the Pareto graphs (right column), with a schematic representation of the measurement areas to be provided at the top. The error bars in the main effect's plots represent the measurement uncertainty from the dataset of the respective product fingerprint (Table 6).

Table 6. Measurement uncertainty of the main effects (U_{me}) per parameter level.

		U_{me}. per Run-Cavity 1							
		Tm [°C]		**Tmld [°C]**		**InjSp [mm/s]**		**PackPr [bar]**	
Position	Unit	220	260	40	60	100	140	440	540
PP1	[μm]	5.6	3.6	3.9	3.8	3.0	3.8	3.2	3.0
PP2	[μm]	5.7	2.2	5.1	4.2	6.2	0.8	3.5	5.7
PP3	[μm]	16.0	1.8	13.3	13.3	17.3	3.2	12.6	14.0
RP2	[μm]	27.5	21.5	23.9	25.6	26.2	21.0	27.2	21.8
		U_{me}. per Run-Cavity 2							
		Tm [°C]		**Tmld [°C]**		**InjSp [mm/s]**		**PackPr [bar]**	
Position	Unit	220	260	40	60	100	140	440	540
PP1	[μm]	3.1	3.2	3.2	3.0	3.1	3.2	3.2	3.1
PP2	[μm]	3.7	4.3	3.9	4.1	3.9	4.2	4.0	3.9
PP3	[μm]	3.7	3.2	3.1	3.5	3.4	3.3	3.4	3.3
RP2	[μm]	46.8	21.8	37.3	39.4	32.8	3.7	35.2	41.3

In comparison to positions PP1, PP2, and PP3 that are located on the molded part, the μ-pillar features in positions RP2 (at the middle of the runner for both Cavities 1 and 2) (Figure 14d) are less sensitive to process variation than the three previously discussed measurand positions. In the case of C2RP2 (Cavity 2—position RP2) a level increase in the Tm, Tmld, InjSp, and PackPr parameters results to a feature height deviation decrease of 8.5 μm, 3.6 μm, 25.2 μm, and 0.66 μm, respectively, revealing the influence of the InjSp parameter. In particular, the Pareto chart in Figure 14d presents the larger influence of InjSp to the measurand C2RP2 in comparison to C1RP2, which is directly connected to the different geometries in Cavities 1 and 2. However, similarly for the results of the feature height deviation from positions C1PP3 and C2PP3 (Figure 14c), none of the parameters' effects can be considered significant due to the overlapping of the uncertainty bars in the presented main effects. The reason for the influence of InjSp and Tm lies again in the lower viscosity of the melt. The melt viscosity in combination with the geometry of μ-structure features, has an effect on the replication of the μ-features, as molten polymer at higher injection speeds (InjSp), or melt temperature (Tm) has a lower value of viscosity and can fill the features before a surface frozen layer is formed. When the packing pressure (PackPr) is considered alone, the already formed frozen layer of the polymer cannot be deformed by the higher packing pressure in order to fill the high aspect ratio μ-pillars. From the main effect plots charts, it can be seen that lower height deviation (i.e., better replication) existed mainly at the positions in the middle of the parts and farther from the gate where the response were less sensitive to process variation. Table 6 presents the measurement uncertainty levels of the main effects shown in Figure 14.

The μ-pillars in position RP2 are sensitive to process variation (although less than the rest of the measurands), and are thus considered suitable "product fingerprint" candidates. The analysis of the effects for the IM process parameters on the eight measurands has provided some indications on the most suitable possible product fingerprints with respect to their sensitivity to process variation. A product fingerprint though is required to have a high level of correlation with the overall part quality assessed by a measurand. In the current concept, the on runner μ-pillars viability as "product fingerprints" is examined.

Thus, the other μ-pillar positions are disregarded since they resulted in non-suitable product fingerprints. A correlation analysis was carried out to determine the most suitable product fingerprint related to the quality of the on part measurands and from the two on runner measurands. For the analysis, the Pearson correlation ρ coefficient was calculated with the use of Equation (16) [27].

$$\rho(x, y) = \frac{\sum_{i=1}^{n}(x_i - \bar{x})(y_i - \bar{y})}{\sqrt{\sum_{i=1}^{n}(x_i - \bar{x})^2 \sum_{i=1}^{n}(y_i - \bar{y})^2}} \tag{16}$$

n is the sample size of the two datasets X and Y ($n*1$ vectors), x_i and y_i: data points in the vectors; $\bar{x}$ and $\bar{y}$: the sample means of datasets X and Y.

The coefficient ρ can vary between -1 and 1, where -1 indicates a perfect negative correlation and 1 indicates a perfect positive correlation. Instead, a ρ value equal to 0 connotes that no correlation exists between the two compared datasets. In this analysis, all the data points from the three replicates of each treatment of a $2^4 \times 3$ full factorial experiment were used for the correlation analysis and the calculation of the absolute Pearson coefficients.

The calculated $|\rho|$ values for the 32 dataset combinations (16 combinations per cavity) are presented in Figure 14. High correlations exist for many dataset combinations, though special focus was given in the correlations of the datasets to the dataset originating to positions RP2 from Cavities 1 (C1RP2) and 2 (C2RP2). Figure 15a is focused on Cavity 1 and it illustrates that the combination dataset with the highest correlation is C1PP1/C1RP2 ($|\rho| = 0.73$) (i.e., near the gate/on the runner), followed by C1PP2/C1RP2 ($|\rho| = 0.60$) and C1PP3/C1RP2 ($|\rho| = 0.57$) (i.e., far from the gate/on the runner), which present a strong correlation for the first combination and moderate correlation for the two consequent ones. Instead, in Cavity 2 no strong correlations exist to the measurands in the cavity, indicating that even though measurand C2RP2 is sensitive to process variations, particularly for injection speed, it is not considered suitable for the quality monitoring of the μ-pillars inside the cavity. Taking into consideration the sensitivity and correlation analyses from measurands in both cavities, only the μ-pillars on the runner of Cavity 1 (C1RP2) can be considered as suitable "product fingerprints" candidate and only for the measurands of Cavity 1.

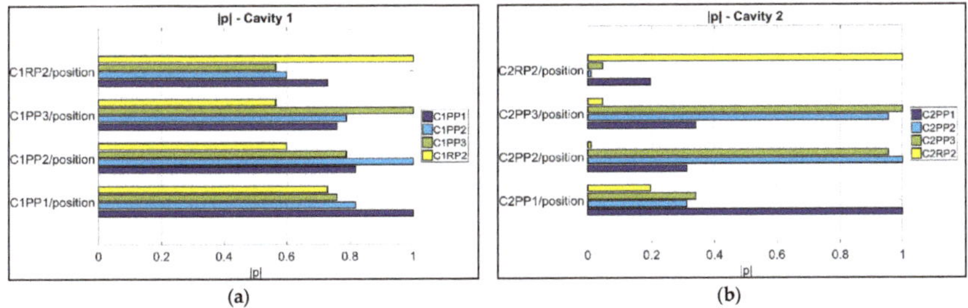

Figure 15. Pearson correlation coefficient plots of measurands to the pillar "product fingerprint" positioned on the runner of the molding (**a**) in Cavity 1 and (**b**) in Cavity 2. A perfect correlation $|\rho| = 1$ exists only for combinations of the same dataset.

3.3. Process Fingerprint Analysis

In the same way as for the "product fingerprint", a set of "process fingerprint" candidates were extracted from the machine process monitoring and regulation signals. The goal was to verify which can act as indicators of the overall product quality, especially for the functional μ-pillar features.

The time-resolved machine data were used to extract two type of indicators:

(1) The first type is characterized by those indicators which originated from the deviation of both the transient injection pressure and injection speed signals with respect to the reference signals such as error of alignment ($\varepsilon(t)$), integrated squared error (ISE), cross correlation shift error (Shift), and dynamic time warping (WarpDis);

(2) Those indicators where the "signal integral" (I_x) and "signal power" (SP_x) were calculated from each signal to extract the information from the signal curve and are subsequently converted into a single value representative of the second type.

3.3.1. Process Fingerprint Based on Indicators of Type 1

As already discussed in Section 2.5.1, the machine controller records the injection speed and the pressure time series and transient data during the process for every consecutive cycle. The deviation of those signals from the initial reference (1st signal per run) is used for the calculation of the deviation-based "process fingerprints" (Type 1). When this type of indicator is considered, the "fingerprints" as well as the dataset's values are standardized in order to be compared. Figure 16 provides an example of the trends that exist between the standardized mold-part deviation measurement and the standardized "process fingerprints" candidate values. It can be seen, particularly for experimental run 16, that not all deviation datasets follow the same trend of the "process fingerprints" candidates. However, the same fingerprints and dataset trends exist for both the nominal (see Figure 16 top) and cross correlated aligned signals (see Figure 16 bottom). It can be seen that "Workdev-InjPr" and "Erwork-InjPr" follow the exact trend with the dataset "C2PP2". Analogously "ISE-InjPr" follows a similar trend. Moreover, the dataset of position RP2 in Cavity 2 (C2RP2) follows a similar trend to fingerprint candidate "ShiftXcorr-InjPr". A similar trend can be observed between the "process fingerprint" candidate "WarpDis-InjPr" and the dataset of position C1PP3.

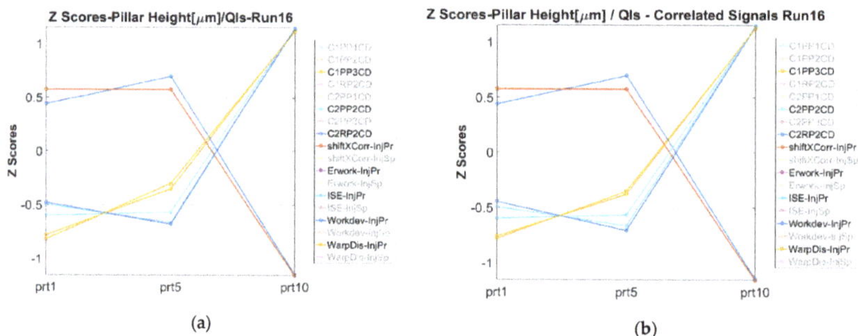

Figure 16. Example of process fingerprint candidates to measurand trends for experimental run 16 based on (**a**) nominal signals and (**b**) cross correlated signals. The legend of the graphs denotes both the measurand datasets (i.e., C1PP1: Cavity 1–Position PP1) and the deviation based (Type 1) "process fingerprints".

When the whole experimental space is considered, the same dataset trends were not always in agreement with the trends of the same candidates. Figure 17 illustrates the occurrence of similar "process fingerprint" trends to the measurement datasets of each experimental run. For example, significant trends between the measurement datasets and the candidate "WarpDis-InjPr" occur a maximum of six times (i.e., six datasets) for Runs 7 and 15 where the Tmelt parameter is kept on the low level. The second process fingerprint candidate occurrence is "Erwork-InjSp" with five times for Run 9 and four times for Run 1. Instead, "process fingerprint" candidates such as "Workdev-InjPr", "ISE-InjPr", and "ISE-InjSp" have less similar trends to the measurement datasets from the same run, even though they appear to have similar trends to measurement datasets from most of the experimental runs.

Run	Tm [°C]	Tmld [°C]	Inj.Sp [mm/s]	Pack.Pr [bar]
R1	220	40	100	440
R2	260	40	100	440
R3	220	60	100	440
R4	260	60	100	440
R5	220	40	140	440
R6	260	40	140	440
R7	220	60	140	440
R8	260	60	140	440
R9	220	40	100	540
R10	260	40	100	540
R11	220	60	100	540
R12	260	60	100	540
R13	220	40	140	540
R14	260	40	140	540
R15	220	60	140	540
R16	260	60	140	540

Process Fingerprint Occurence

Run	shiftXCorr-InjPr	shiftXCorr-InjSp	Erwork-InjPr	Erwork-InjSp	ISE-InjPr	ISE-InjSp	Workdev-InjPr	Workdev-InjSp	WarpDis-InjPr	WarpDis-InjSp
R16	3	1	0	2	2	1	0	0	2	0
R15	0	1	0	0	3	3	1	0	6	2
R14	1	1	0	2	0	3	2	0	1	2
R13	3	2	0	1	0	1	2	0	0	0
R12	2	0	0	2	2	0	1	0	1	0
R11	1	1	1	1	1	2	1	0	3	1
R10	0	1	0	1	1	0	1	0	1	0
R9	2	2	0	5	2	2	2	0	2	0
R8	2	2	0	2	2	1	0	0	1	1
R7	0	1	0	0	3	3	1	0	6	2
R6	1	1	0	2	0	3	2	0	1	2
R5	3	2	0	1	0	1	2	0	0	0
R4	2	1	0	2	2	0	1	0	1	2
R3	0	1	0	1	2	2	0	0	2	1
R2	0	1	0	2	0	0	1	0	0	0
R1	3	2	0	4	1	2	2	1	1	0

Figure 17. Process fingerprint candidate trend occurrence per Run.

As a conclusion, "process fingerprints" "Workdev-InjPr", "ISE-InjPr", and "ISE-InjSp" together with "WarpDis-InjPr" are considered suitable for the quality control of the pillar μ-features in most of the examined experimental space. However, their correlation and trend are directly dependent on each of the treatments' process parameter combination.

3.3.2. Process Fingerprint Based on Indicators of Type 2

The second type of "process fingerprint" candidates originates from each signal individually. To examine the suitability of signal integrals and signal power to serve as "process fingerprint" candidates, a correlation analysis to respective measurement datasets was conducted with the correlation coefficients $|\rho|$ to be presented in Figure 18a for Cavity 1 and Figure 18b for Cavity 2. The maximum correlation coefficient ($|\rho|$ = 0.436, indicating a moderate correlation) values occur for the combination I.InjSp/C2PP2 (integral of injection speed signal vs. the dataset in position C2PP2, in the middle of the part). The rest of the combinations had weak correlation: they exhibited $|\rho|$ values lower than 0.4. For this reason, the integral and power of the injection pressure and speed signals originating from the IM machine were not considered suitable "process fingerprint" candidates for the quality control and assurance for μ-pillar structured molded components for the particular application.

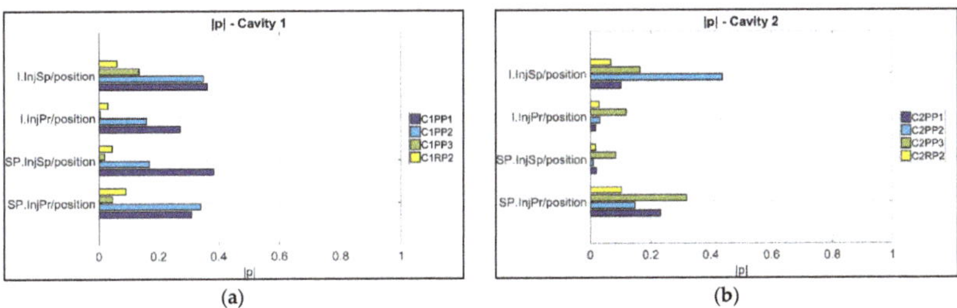

Figure 18. Pearson correlation coefficient plots of measurands to the pillar "product fingerprint" positioned on the runner of the molding (**a**) in Cavity 1 and (**b**) in Cavity 2.

4. Conclusions

A new approach towards process monitoring and fast-integrated quality assurance of injection molded microstructured components based on product and process fingerprints was presented and validated in this paper. The concept is examined on two parallel tracks. Micro pillars were positioned on the runner before each cavity to serve as "product fingerprints" and the process controlling signals were collected to extract "process fingerprint" candidates. The suitable fingerprints were selected after a sensitivity and correlation analysis was conducted to assess their sensitivity to process variation and correlation, respectively. As far as the product quality assurance was concerned, the replication quality of the μ-pillars was assessed using 3D scanning focus variation microscopy (i.e., off-line metrology). For the process monitoring, the signals generated by the machine regulation embedded sensors were used to extract the time-resolved data. Summarizing the key findings of the research, the following conclusions can be drawn:

- The variation of the IM process parameters settings has an effect on the manufacturing quality and replication of the molded μ-pillar structured components placed both in the cavities as well as on the runners.

- The variation of the process was used to assess the suitability of μ-pillars in the eight different positions to act as product fingerprint. The analysis was based on their replication quality. A correlation analysis was then used for verification. This track was focused on the μ-pillars positioned on the runners of the molding in positions C1RP2 and C2RP2. For Cavity 1, it can be seen that the dataset position C1RP2 can be used to monitor the quality of the μ-features on the part, especially for position C1PP1 (near the gate) with the highest correlation to originate to the combination is C1PP1/C1RP2 ($|\rho|$ = 0.73) (i.e., near the gate/on the runner), followed by C1PP2/C1RP2 ($|\rho|$ = 0.60) and C1PP3/C1RP2 ($|\rho|$ = 0.57) (i.e., far from the gate/on the runner). Instead, the μ-pillars on the runner of Cavity 2 (C2RP2), did not present strong correlations with respect to the measurands of the features in the cavity, indicating that these μ-pillars are not suitable to serve as a "product fingerprint".

- Two different types of process fingerprint candidates were assessed for their suitability to act as quality indicators of the micro structures on the molded parts. Results show that only a small number of process fingerprint candidates from the category of deviation-based process fingerprints (i.e., Type 1) were considered suitable for process monitoring when considered together with the proper measurand. From the Type 2 indicators in fact, no candidate presented a strong correlation with the quality of a measurand. This indicates that the integral and signal power of machine injection pressure and speed signals could not be used for the monitoring of the overall part quality in the current application.

- Finally, it can be concluded that the deviation of the quality of the part's μ-pillars can be monitored by monitoring the deviation of the "Workdev-InjPr", "ISE-InjPr", "ISE-InjSp", and "WarpDis-InjPr" process fingerprints. These fingerprints present similar trends with measurands for most of the treatments in the investigated process window.

Future work will aim at the validation of the presented concept, enriched with data acquired from in-mold temperature and pressure sensors. Furthermore, the assessment of product and process fingerprints performance robustness will be carried out in longer manufacturing runs emulating an actual production environment.

Author Contributions: N.G., G.T., and Y.Z. conceived and designed the experiments; N.G. performed the experiments and measurements; N.G. analyzed the data; G.T. and Y.Z. consulted on the work; N.G. wrote the paper; G.T. and Y.Z. revised the paper.

Funding: This research was funded by Innovation Fund Denmark grant number 3067-00001B and the APC was funded by the Technical University of Denmark.

Acknowledgments: This paper reports work undertaken within the framework of the project MADE (Manufacturing Academy of Denmark, http://en.made.dk/) Work Package 3 "3D Print and New Production Processes" of the Research Platform MADE SPIR (Strategic Platform for Innovation and Research, http://en.made.dk/spir/). MADE is a collaborative research project supported both by the Danish Manufacturing Industry and by the Innovation Fund Denmark (https://innovationsfonden.dk/en). MADE and Innovation Fund Denmark are thanked for providing financial support to the PhD project "Precision Injection Moulding of Micro Features using Integrated Process/Product Quality Assurance".

Conflicts of Interest: The authors declare no conflict of interest.

References

1. Alting, L.; Kimura, F.; Hansen, H.N.; Bissacco, G. Micro Engineering. *CIRP Ann. Manuf. Technol.* **2003**, *52*, 635–657. [CrossRef]
2. Brousseau, E.B.; Dimov, S.S.; Pham, D.T. Some recent advances in multi-material micro- and nano-manufacturing. *Int. J. Adv. Manuf. Technol.* **2010**, *47*, 161–180. [CrossRef]
3. Wang, X.; Gu, J.; Shen, C.; Wang, X. Warpage optimization with dynamic injection molding technology and sequential optimization method. *Int. J. Adv. Manuf. Technol.* **2015**, *78*, 177–187. [CrossRef]
4. Theilade, U.A.; Hansen, H.N. Surface microstructure replication in injection molding. *Int. J. Adv. Manuf. Technol.* **2006**, *33*, 157–166. [CrossRef]
5. Giannekas, N.; Tosello, G.; Zhang, Y. A study on replication and quality correlation of on-part and on-runner polymer injection molded micro features. In Proceedings of the World Congress on Micro and Nano Manufacturing (WCMNM), Kaohsiung, Taiwan, 27–30 March 2017; pp. 365–368.
6. Marhöfer, D.M.; Tosello, G.; Islam, A.; Hansen, H.N. Gate Design in Injection Molding of Microfluidic Compoments Using Process Simulations. *J. Micro Nano-Manuf.* **2016**, *4*, 025001. [CrossRef]
7. Tosello, G.; Gava, A.; Hansen, H.N.; Lucchetta, G. Study of process parameters effect on the filling phase of micro-injection moulding using weld lines as flow markers. *Int. J. Adv. Manuf. Technol.* **2010**, *47*, 81–97. [CrossRef]
8. Calaon, M. Process Chain Validation in Micro and Nano Replication. Ph.D. Thesis, Technical University of Denmark, Kongens Lyngby, Denmark, 2014.
9. Gao, R.X.; Tang, X.; Gordon, G.; Kazmer, D.O. Online product quality monitoring through in-process measurement. *CIRP Ann. Manuf. Technol.* **2014**, *63*, 493–496. [CrossRef]
10. Kusić, D.; Kek, T.; Slabe, J.M.; Svečko, R.; Grum, J. The impact of process parameters on test specimen deviations and their correlation with AE signals captured during the injection moulding cycle. *Polym. Test.* **2013**, *32*, 583–593. [CrossRef]
11. Wong, H.Y.; Fung, K.T.; Gao, F. Development of a transducer for in-line and through cycle monitoring of key process and quality variables in injection molding. *Sens. Actuators A* **2008**, *141*, 712–722. [CrossRef]
12. Kazmer, D.O.; Gordon, G.W.; Mendible, G.A.; Johnston, S.P.; Tang, X.; Fan, Z.; Gao, R.X. A Multivariate Sensor for Intelligent Polymer Processing. *IEEE/ASME Trans. Mechatron.* **2014**, *1*, 1–9. [CrossRef]
13. Mendibil, X.; Llanos, I.; Urreta, H.; Quintana, I. In process quality control on micro-injection moulding: The role of sensor location. *Int. J. Adv. Manuf. Technol.* **2017**, *89*, 3429–3438. [CrossRef]
14. Chen, Z.; Turng, L.-S.; Wang, K.-K. Adaptive Online Quality Control for Injection Molding by Monitoring and Controling Mold Separation. *Polym. Eng. Sci.* **2006**, *46*, 569–580. [CrossRef]
15. Johnston, S.; Mccready, C.; Hazen, D.; Vanderwalker, D.; Kazmer, D. On-Line Multivariate Optimization of Injection Molding. *Polym. Eng. Sci.* **2015**, *55*, 1–8. [CrossRef]
16. Yang, Y.; Yang, B.; Zhu, S.; Chen, X. Online quality optimization of the injection molding process via digital image processing and model-free optimization. *J. Mater. Process. Technol.* **2015**, *226*, 85–98. [CrossRef]
17. Ozcelik, B.; Erzurumlu, T. Comparison of the warpage optimization in the plastic injection molding using ANOVA, neural network model and genetic algorithm. *J. Mater. Process. Technol.* **2006**, *171*, 437–445. [CrossRef]
18. Liu, C.; Manzione, L.T. Process Studies in Precision Injection Moulding I: Process Parameters and Precision. *Polym. Eng. Sci.* **1996**, *36*, 1–9. [CrossRef]
19. *Moldflow Synergy*; Autodesk: San Rafael, CA, USA, 2017.
20. Dansk Standards-ISO. 15530-3 Geometrical Product Specifications (GPS)—Coordinate Measuring Machines (CMM)—Technique for Determining the Uncertainty of Measurement—Part 3: Use of Calibrated Workpieces or Measurements Standards. 2011. Available online: https://www.ds.dk/en/standards/buy-standards (accessed on 8 October 2018).
21. Gasparin, S.; Hansen, H.N.; Tosello, G. Traceable surface characterization using replica moulding technology. In Proceedings of the 13th International Conference on Metrology and Properties of Engineering Surfaces, Lyngby, Denmark, 12–15 April 2011; pp. 310–315.

22. Giannekas, N.; Gammelby, R.; Tosello, G.; Tcherniak, D.; Zhang, Y. Feasibility study on integrated process/product quality assurance framework for precision injection moulding based on vibration monitoring. In Proceedings of the Euspen's 17th International Conference & Exhibition, Hannover, Germany, 29 May–2 June 2017; pp. 508–509.

23. Gajic, Z. Correlation of Discrete-Time Signals. In *Linear Dynamic Systems and Signals*; Prentice Hall: Englewood Cliffs, NJ, USA, 2003; pp. 90–100.

24. Chen, H.; Li, Y.; Wu, Z. Dynamic time warping distance method for similarity test of multipoint ground motion field. *Math. Probl. Eng.* **2010**, *2010*. [CrossRef]

25. Sakoe, H.; Chiba, S. Dynamic Programming Algorithm Optimization for Spoken Word Recognition. *IEEE Trans. Acoust.* **1978**, *26*, 43–49. [CrossRef]

26. Sanguansat, P. Multiple Multidimensional Sequence Alignment Using Generalized Dynamic Time Warping. *WSEAS Trans. Math.* **2012**, *11*, 668–678.

27. Lee Rodgers, J.; Nicewander, W.A. Thirteen ways to look at the correlation coefficient. *Am. Stat.* **1988**, *42*, 59–66. [CrossRef]

Journal of
*Manufacturing and
Materials Processing*

Article

Machining Forces Due to Turning of Bimetallic Objects Made of Aluminum, Titanium, Cast Iron, and Mild/Stainless Steel

AMM Sharif Ullah

Faculty of Engineering, Kitami Institute of Technology, Kitami 090-8507, Japan; ullah@mail.kitami-it.ac.jp

Received: 28 August 2018; Accepted: 10 October 2018; Published: 11 October 2018

Abstract: This article elucidates the characteristics of machining forces (an important phenomenon by which machining is studied) using three sets of bimetallic specimens made of aluminum–titanium, aluminum–cast iron, and stainless steel–mild steel. The cutting, feed, and thrust forces were recorded for different cutting conditions (i.e., different cutting speeds, feeds, and cutting directions). Possibility distributions were used to quantify the uncertainty associated with machining forces, which were helpful in identifying the optimal machining direction. In synopsis, it was found that while machining the steel-based bimetallic specimens, keeping a low feed and high cutting speed is the better option, and the machining operation can be performed in both the hard-to-soft and soft-to-hard material directions, but machining in the soft-to-hard material direction is the better option. On the other hand, very soft materials should not be used in fabricating a bimetallic part because it creates machining problems. Cutting power was estimated using the cutting and feed force signals. Manufacturers who support sustainable product development (including design, manufacturing, and assembly) can benefit from the outcomes of this study because parts/products made of dissimilar materials (or multi-material objects) are better than their mono-material counterparts in terms of sustainability (cost, weight, and CO_2 footprint).

Keywords: sustainability; bimetallic object; cutting force; uncertainty; machining power

---

1. Introduction

The research on machining is mostly concerned with the machining of objects made of mono-material and special alloys. On the other hand, research on the machining of objects made of multiple materials cannot be ignored, mainly because of the rising concerns for sustainability. The explanation is given below.

In general, sustainability means fulfilling the present generation's needs without compromising the ability to fulfill the future generations' needs [1]. In more specific terms, sustainability means ensuring material efficiency, energy efficiency, and component efficiency, preferably simultaneously, for all products that inhabit the artificial world [2]. Here, material efficiency is with respect to the usages of materials and takes into account the issues regarding energy consumption and resource depletion while producing the primary materials; it also considers issues like the cost and weight reduction of a product [2–5]. Energy efficiency takes into account the energy consumption during the manufacturing activities (e.g., machining and assembly) of a product [2,5]. Component efficiency takes into account the degree of fulfillment of the intended functionality, quality, and reliability requirements of the components used in a product [2]. The interplay of these efficiencies is presented in detail in [2], where it is concluded that material efficiency is more effective than the other two efficiencies in enhancing the sustainability of a product. For example, a multi-material object is better than its monometallic counterpart (e.g., an object made of aluminum–titanium is better than its monometallic counterpart made of Titanium only, in terms of cost, weight, and CO_2 footprint) [2]. Increasing the material efficiency might affect the energy and component efficiencies, which is not desirable. Therefore, optimization is needed to obtain the best that a multi-material object can offer.

Nevertheless, the usages of multi-material products are expected to increase in the years to come due to the fact mentioned above (i.e., enhancing the sustainability of a product from the viewpoint of material efficiency). Nowadays, both physical joining processes (e.g., friction welding) [6–9] and additive manufacturing processes (e.g., selective laser sintering) [10–13] are used to manufacture objects made of dissimilar metals. The advent of such manufacturing processes will also accelerate the usages of multi-material products since these processes help manufacture different parts made of different types of dissimilar metals. It is worth mentioning that additive manufacturing processes that add materials layer by layer based on the solid model of an object have been found suitable for manufacturing very complex and highly customized objects using multiple materials [10–13]. As such, additive manufacturing processes (selective laser sintering) can easily fabricate an object made of multiple materials, which is often difficult to achieve by conventional manufacturing processes (e.g., machining, casting, forming, and welding).

The above explanation refers to the fact that more and more objects made of multiple materials will inhabit our surroundings in the years to come. However, a multi-material object manufactured either by additive manufacturing or by other manufacturing processes (e.g., friction welding) must be machined so that it achieves the required dimensional accuracy and surface finish. This necessitates machining knowledge regarding multi-material objects. In the literature, a relatively limited number of studies are found regarding the machining of objects made of dissimilar materials. In particular, the studies reported in [14–21] are noted. These studies show that the machining of a multi-material object entails some unique properties. For example, a monometallic workpiece can be machined from any sides, whereas while machining a workpiece made of two different materials, the machining direction must be optimized (e.g., machining from the softer material side to the harder material side or vice versa) [20]. The surface roughness quantification process of an object made of two different metals needs some unconventional parameters (e.g., entropy, possibility distribution, and the like) [19,21]. The main issue of such uniqueness is the existence of the joint area or heat-affected zone, where the material compositions and properties (particularly hardness) exhibit a great deal of variability compared to the constituent materials. The authors in [6–9,22] have described this issue elaborately. Depending on whether a cutting tool passes the joint area from the softer material side to the harder material side, or vice versa, the machining characteristics might differ. As a result, the machining forces (cutting force, feed force, and so on) might exhibit a different kind of character when the cutting tool passes the joint area either from the softer material side to the harder material side or vice versa. Since machining forces provide valuable insights into machining phenomena [23], it is worth investigating the nature of the machining forces that arise when a cutting tool passes the joint area from both sides of a bimetallic specimen. From this contemplation, this article reports the characteristics of machining forces that occur when turning three sets of dissimilar metallic specimens made of aluminum–titanium, aluminum–cast iron, and stainless steel–mild steel. Accordingly, the remainder of this article is organized as follows. Section 2 describes the bimetallic specimens, experimental setup, and data acquisition technique. Section 3 presents the characteristics of the machining forces underlying the stainless steel–mild steel in terms of time series data and uncertainty. Section 4 presents the characteristics of the machining forces underlying the aluminum–titanium in terms of time series data and uncertainty. Section 5 presents the characteristics of the machining forces underlying the aluminum–cast iron in terms of time series data and uncertainty. Section 6 discusses the implication of the results. Section 7 provides the concluding remarks of this study.

2. Machining Experiments and Data Acquisition

This section describes the bimetallic specimens, experimental setup, and data acquisition technique used while turning the bimetallic specimens.

Three different sets of bimetallic specimens were fabricated using friction welding [6,7]. The description of the welding conditions can be found in [2]. Table 1 lists the materials used to prepare the specimens. The tensile strength, percent elongation, and hardness of each material are also listed in Table 1.

Table 1. Materials used for fabricating the dissimilar metallic specimens.

Bimetallic Specimens	Materials	Tensile Strength (MPa)	Elongation (%)	Hardness (Scale)
SU–SC	Stainless Steel (JIS: SUS304)	663	55	182 (HV)
	Mild Steel (JIS: S15CK)	439	38	132 (HV)
Al–Ti	Aluminum (JIS: A1070)	120	27	41 (HV)
	Commercial Pure (CP) Titanium	401	35	146 (HV)
Al–CI	Aluminum (JIS: A5052)	265	17.4	86 (HV)
	Ductile Cast Iron	442	18.6	79.2 (HRB)

The first set of specimens, defined as SU–SC, was prepared by joining two different materials, namely, stainless steel (JIS: SUS304) and mild steel (JIS: S15CK). The chemical composition (wt%) of the stainless steel was as follows: 0.052 C, 0.416 Si, 1.529 Mn, 0.0319 P, 0.0186 S, 8.057 Ni, 18.293 Cr, 0.185 Mo, 0.483 Cu, and 70.9345 Fe. The chemical composition (wt%) of the mild steel was as follows: 0.15 C, 0.20 Si, 0.40 Mn, 0.19 P, 0.022 S, 0.03 Ni, 0.14 Cr, 0.02 Cu, and 98.848 Fe. The tensile strength (i.e., ultimate strength), elongation, and hardness of the stainless steel were 663 MPa, 55%, and 182 HV, respectively. The tensile strength (i.e., ultimate strength), elongation, and hardness of the Mild Steel were 439 MPa, 38%, and 132 HV, respectively. The second set of specimens, defined as Al–Ti, was prepared by joining two different materials, namely, aluminum (JIS: A1070) and commercial pure (CP) titanium. The chemical composition (wt%) of the aluminum (JIS: A1070) were as follows: 0.03 Si, 0.10 Fe, 0.01 Cu, 0.02 Mg, 0.01 V, 0.01 Ti, others $\leq$ 0.03 others, and 99.82 Al. The chemical composition (wt%) of the CP titanium was as follows: 0.0011 H, 0.089 O, 0.006 N, 0.038 Fe, 0.005 C, and 99.8609 Ti. The tensile strength (i.e., ultimate strength), elongation, and hardness of the aluminum (JIS: A1070) were 120 MPa, 27%, and 41 HV, respectively. The tensile strength (i.e., ultimate strength), elongation, and hardness of the CP titanium were 401 MPa, 35%, and 146 HV, respectively. The other set of specimens, defined as Al–CI, was prepared by joining two different materials, namely, aluminum (JIS: A5052) and ductile cast iron. The chemical composition (wt%) of the aluminum (JIS: A5052) was as follows: 0.09 Si, 0.16 Fe, 0.02 Cu, 0.03 Mn, 2.6 Mg, 0.25 Cr, 0.01 Zn, $\leq$0.15 others, and 96.69 Al. The chemical composition (wt%) of the ductile cast iron was as follows: 3.76 C, 2.91 Si, 0.49 P, 0.011 S, 0.029 Mg, and 92.8 Fe. The tensile strength (i.e., ultimate strength), elongation, and hardness of the aluminum (JIS: A5052) were 265 MPa, 17.4%, and 86 HV, respectively. The tensile strength (i.e., ultimate strength), elongation, and hardness of the ductile cast iron were 442 MPa, 18.7%, 79.2 HRB, respectively.

Note that the tensile strength, percent elongation, and hardness of one of the constituent materials are greater than those of the other for each set of specimens. This ensures machining of soft-to-hard material or vice versa at the joint area. Figure 1 shows the pictures of the specimens, one from each set of specimens. The flash generated in the joint area (see Figure 1) was removed by using a turning operation before conducting the machining experiments for obtaining the machining force data. The friction welding conditions used to prepare the bimetallic specimens (Figure 2) are listed in Table 2. As seen in Table 2, for the specimens called SU-SC, the rotating material was S15CK (i.e., mild steel). For the specimens called Al-Ti, the rotating material was A1070 (i.e., aluminum). For the other specimens, the rotating material was A5052 (aluminum). The diameters of rotating material (while performing friction welding) for all specimens were 12 mm. The friction speed, friction pressure, and upset time were 27.5 s^{-1} (1650 rpm), 30 MPa, and 6 s, respectively, for all specimens. Whereas, the friction times for the specimens namely SU-SC, Al-Ti, and Al-CI were 2 s, 1 s, and 3 s, respectively. The upset pressures for the specimens, namely SU-SC, Al-Ti, and Al-CI were 270 MPa, 90 MPa, and 200 MPa, respectively.

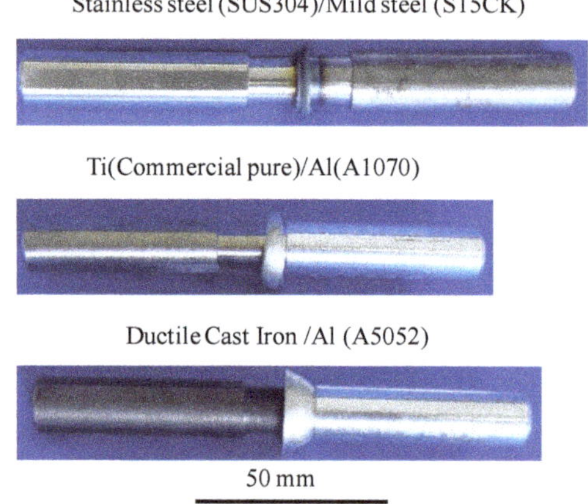

Figure 1. The pictures of the bimetallic specimens.

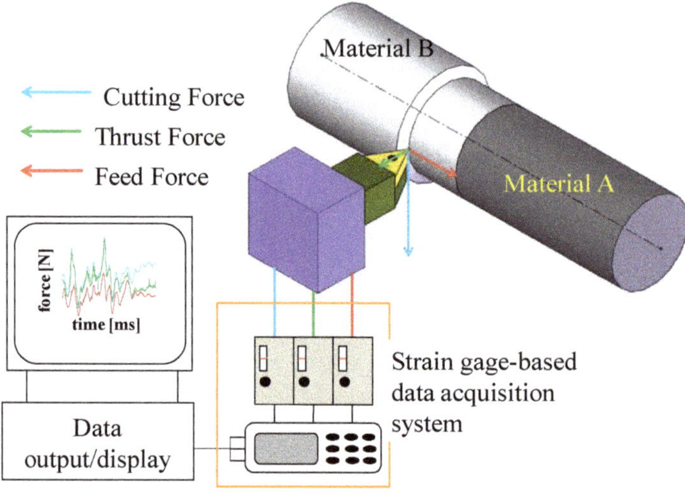

Figure 2. Experimental setup.

On the other hand, the cutting conditions for the machining experiments are summarized in Table 3. Carbide inserts (TNMG160404-MF) supplied by Sandvik^{TM} were used as cutting tools for the machining experiments. Two cutting speeds (v_c), 25 m/min and 50 m/min, were used here. The reason for using such cutting velocities is that most job-shop type workshops, where machining is carried out in real-life settings, are often forced to use very low cutting velocities due to resource constraints: see [24] for a detailed description on the choice of cutting speed based on real-life constraints. However, the rotational speed of the chuck was adjusted in every machining run, ensuring the above cutting velocities. The cutting speeds also ensure no or less tool wear during each machining run. Similar to cutting speed, two values of feed (f), 0.1 mm/rev and 0.2 mm/rev, were used, whereas the depth of cut (a_p) was kept constant (1 mm) for all machining runs. The machining experiments were conducted at three different zones of each specimen: the zones of the constituent materials and the joint area. In Figure 2, one of the constituent materials is denoted as Material A and the other is denoted as Material B. According to Table 1, Material A means stainless steel (JIS: SUS304), aluminum (JIS: A1070), or aluminum (JIS: A5052), for the specimen SU–SC, Al–Ti, or Al–CI, respectively. Similarly, Material B means mild steel (JIS: S15CK), commercial pure (CP) titanium, or ductile cast iron, for the specimen SU–SC, Al–Ti, or Al–CI, respectively.

Table 2. Friction welding conditions for fabricating the dissimilar metallic specimens.

Friction Welding Conditions	Specimens		
	SU-SC	Al-Ti	Al-CI
Rotating material	S15CK	A1070	A5052
Diameter of the rotating material (mm)		12	
Friction speed (s^{-1})		27.5 (1650 rpm)	
Friction pressure (MPa)		30	
Friction time (s)	2	1	3
Upset pressure (MPa)	270	90	200
Upset time (s)		6	

Table 3. Cutting conditions for machining experiments.

Items	Descriptions
Machine Tool	Lathe Machine Make: WASHINO Model: LEO-80A
Cutting Tool	Carbide CVD Coated Insert Make: Sandvik™ Code: TNMG160404-MF
Cutting Speed (v_c) (m/min)	25, 50
Rotational Speeds of the Chuck (rpm)	1377
Feed (f) (mm/rev)	0.1, 0.2
Depth of Cut (a_p) (mm)	1
Cutting Direction	A to B, B to A (for the joint area)

The joint area was machined from both directions—the hard-to-soft material direction and vice versa (i.e., from the Material A to Material B directions, and vice versa)—for each specimen. To do this, the machining force signals for a machining length of about 4 mm were recorded using a strain gage-based data acquisition system, as schematically illustrated in Figure 2. As seen in Figure 2, the system outputs the machining forces from three different channels. One of the channels records the forces in the direction of the cutting speed. The force signals recorded from this channel are called cutting force signals. Another channel records the forces in the direction of the feed. The force signals recorded from this channel are called feed force signals. The other channel records the forces in the direction of the tool post. The force signals recorded from this channel are called thrust force signals. The signals were recorded after every 0.2 ms for the three channels. It is worth mentioning that the cutting and feed force signals were used to calculate the cutting power and thereby to determine the specific cutting energy/pressure. The thrust force signals were not used in the calculations but recorded for the sake of having a complete picture of the machining phenomena.

However, for the sake of analysis, the raw signals require sampling. Figure 3 schematically illustrates the sampling technique. The description is as follows. The time series of the force signals consists of the signals produced when the cutting tool approaches the cutting zone, when the cutting tool is removing materials, and when the cutting tool moves away from the cutting zone. Therefore, the raw signals, as shown in Figure 3a, require sampling. To do the sampling, a sampling span, i.e., a time interval, was chosen in such a way that the signals in the sampling span consist of cutting/feed/thrust force signals only when the cutting tool removes the materials either in the constituent material zone (i.e., in the zone of Material A and Material B) or in the joint area (i.e., the segment where Material A and B are physically connected). The case shown in Figure 3 corresponds to the sampling of the machining force signals in the joint area. The force signals after sampling were reset to a time equal to zero. Thus, the following relationships hold between the raw and sampled signals.

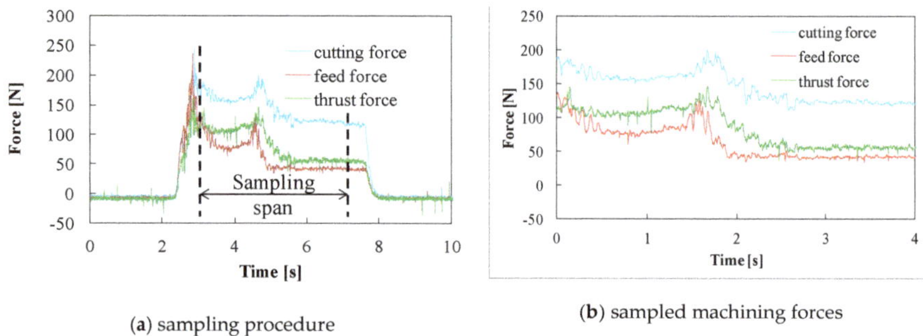

(a) sampling procedure   (b) sampled machining forces

Figure 3. Force data after sampling.

Let $F_{RX}(t)$, $t = 0, \Delta, \ldots, T1, T1 + \Delta, \ldots, T2, \ldots$ be the raw signals of X, $\forall X \in \{C, F, T\}$. Here, C, F, and T mean cutting, feed, and thrust force signals, respectively. The interval $[T1, T2]$ is the sampling span. The symbol Δ is the sampling interval of the raw signals $F_{RX}(t)$. As mentioned before, here $\Delta = 0.2$ ms. The segment of signals $F_{RX}(t = T1), \ldots, F_{RX}(t = T2)$ is used to get the sampled signals. However, the time interval in the sampled signal can be increased for the sake of analysis. Let $F_{SX}(\tau)$ be the sampled signals. Thus, $F_{SX}(\tau = 0) = F_{RX}(t = T1)$, $F_{SX}(\tau = \lambda\Delta) = F_{RX}(t = T1 + \lambda\Delta)$, $\ldots, F_{SX}(\tau = n\lambda\Delta) = F_{RX}(t = T2)$. This means that the sampled signal consists of $n + 1$ data points, and the data points are collected using a time interval $\lambda\Delta$. If $\lambda = 5$, and $\Delta = 0.2$ ms, then $\lambda\Delta = 1$ ms, i.e., the time interval of the sampled signal is 1 ms. Therefore, $F_{SX}(\tau)$ means cutting, feed, or thrust force signals at a time interval of 1 ms where $X = C, F,$ or T, respectively. This convention is used throughout this article. The pictures of the specimens taken after machining are shown in Appendix A.

3. Analyzing Machining Forces Underlying SU–SC

This section describes the machining forces underlying the bimetallic specimens denoted as SU–SC.

Figure 4 shows the machining forces (thrust, feed, and cutting forces) in the time domain. The plots in Figure 4a,d,g,j show the expected machining forces of the constituent materials (S15CK + SUS304), neglecting the joint area. The plots in Figure 4b,e,h,k show the machining forces manifested in the joint area while machining from the S15CK direction to the SUS304 direction. The plots in Figure 4c,f,i,l show the machining forces manifested in the joint area while machining from the SUS304 direction to the S15CK direction. As seen in Figure 4, if a low feed (0.1 mm/rev) and low cutting speed (25 m/min) are used and machining is done from the hard material (SUS304) direction to the soft material (S15CK) direction, then the machining forces can be reduced. When a high feed is preferred, then the choice is to machine from the opposite direction—from the soft material (S15CK) direction to the hard material (SUS304) direction. For a high cutting speed (50 m/min), this argument is still valid for both low and high feeds, but a low feed is perhaps a better option.

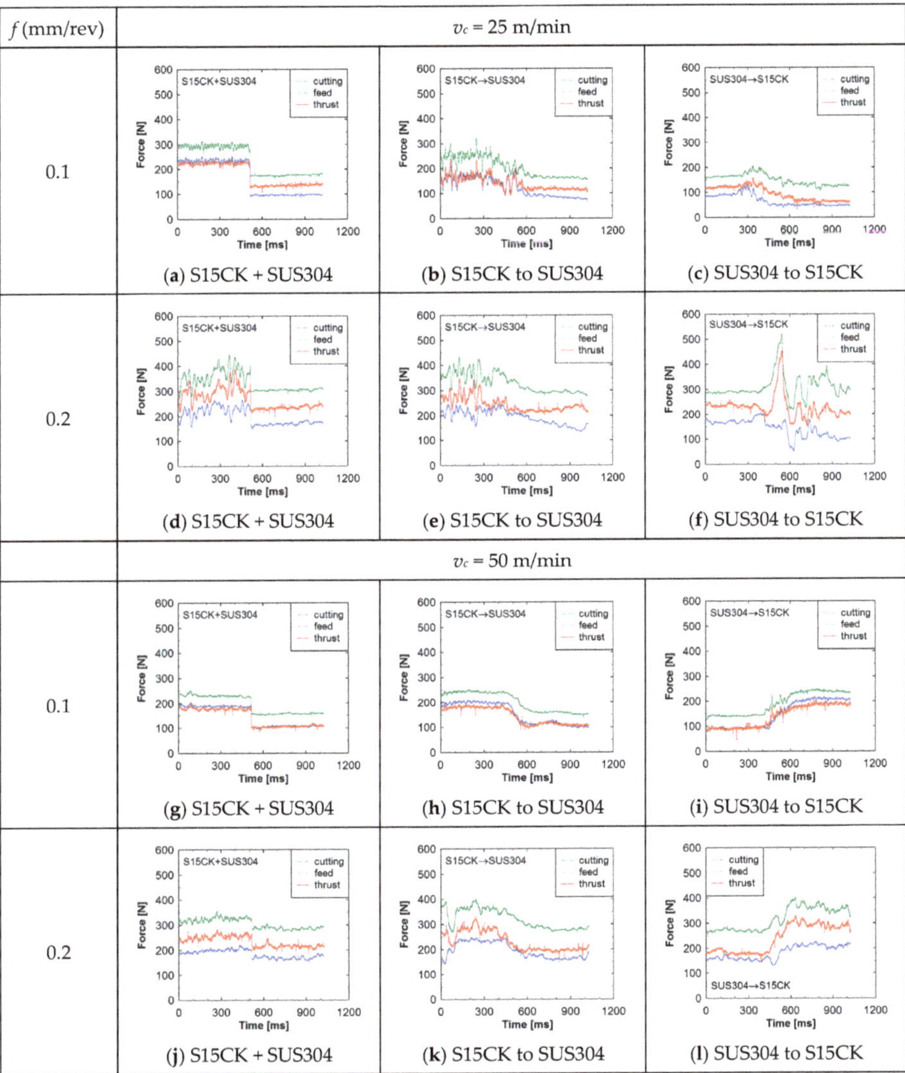

Figure 4. Machining forces underlying SU–SC.

To be more specific, the uncertainty in the cutting forces was studied by constructing the possibility distributions [25,26] (probability-distribution-neutral representation of uncertainty) for the cutting forces, as shown in Figure 4. Appendix B shows the mathematical settings for inducing a possibility distribution from a set of numerical data. The results are shown in Figure 5. In the plots in Figure 5, the phrase "DoB" means the degree of belief (or membership value, see Appendix B), which is a value in the interval [0, 1]. The possibility distributions also support the abovementioned conclusions regarding the relationships between cutting conditions and cutting forces. In particular, the possibility distributions show that the use of a low feed and low cutting speed and the cutting direction hard-to-soft is a better option for reducing the cutting force and its uncertainty.

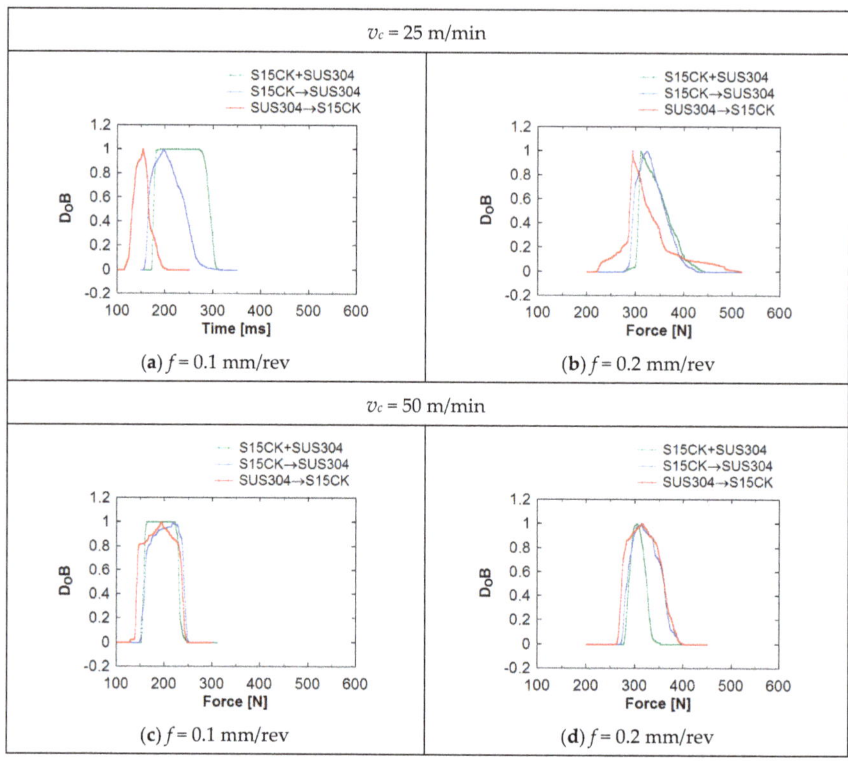

Figure 5. Uncertainties in the machining forces underlying SU–SC.

4. Analyzing Machining Forces Underlying Al–Ti

This section describes the machining forces underlying the bimetallic specimens denoted as Al–Ti. It is worth mentioning that this is a uniform combination similar to SU–SC because the tensile strength, hardness, and percent elongation of CP titanium are greater than those of aluminum (A1070), as listed in Table 1. As such, it will help validate the conclusion made in the previous section.

Figure 6 shows the machining forces (thrust, feed, and cutting forces) in the time domain for the dissimilar metallic specimens denoted as Al–Ti for the cutting conditions listed in Table 1. The plots in Figure 6a,d,g,j show the expected machining forces of the constituent materials (Al + Ti), neglecting the joint area. The plots in Figure 6b,e,h,k show the machining forces manifested in the joint area while machining from the Al direction to the Ti direction. The plots in Figure 6c,f,i,l show the machining forces manifested in the joint area while machining from the Ti direction to the Al direction. As seen in Figure 6, if a low feed (0.1 mm/rev) and low cutting speed (25 m/min) are used and the machining is done from the soft material (Al) direction to the hard material (Ti) direction, then the machining forces can be reduced. The same conclusion regarding the feed is valid for a high cutting speed. This is somewhat an opposing conclusion compared to that of the previous case. The reason for this somewhat dissimilar result is perhaps the hardness of the materials. Here, Al is too soft compared to the other material. This means that when a very soft metal is used in a dissimilar metallic object, it is better to start the machining operation from the soft material side using a low feed and low cutting speed.

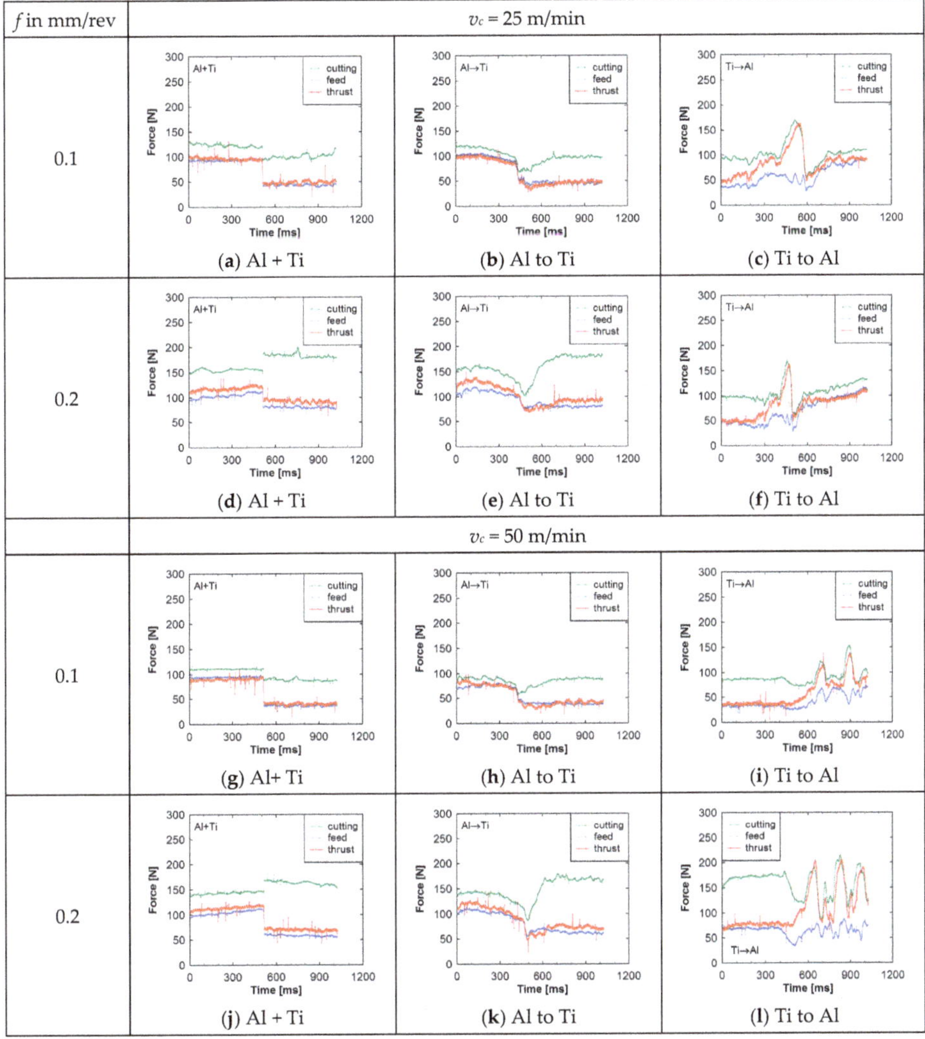

Figure 6. Machining forces underlying Al–Ti.

To be more specific, the uncertainty in the cutting forces shown in Figure 6 was further studied by constructing possibility distributions similar to the previous case. The results are shown in Figure 7. The possibility distributions also support the abovementioned conclusions regarding the relationships between cutting conditions and cutting forces. In particular, the possibility distribution shows that the use of a low feed and low cutting speed and employing the cutting direction soft-to-hard is the right approach for reducing the cutting force and its uncertainty.

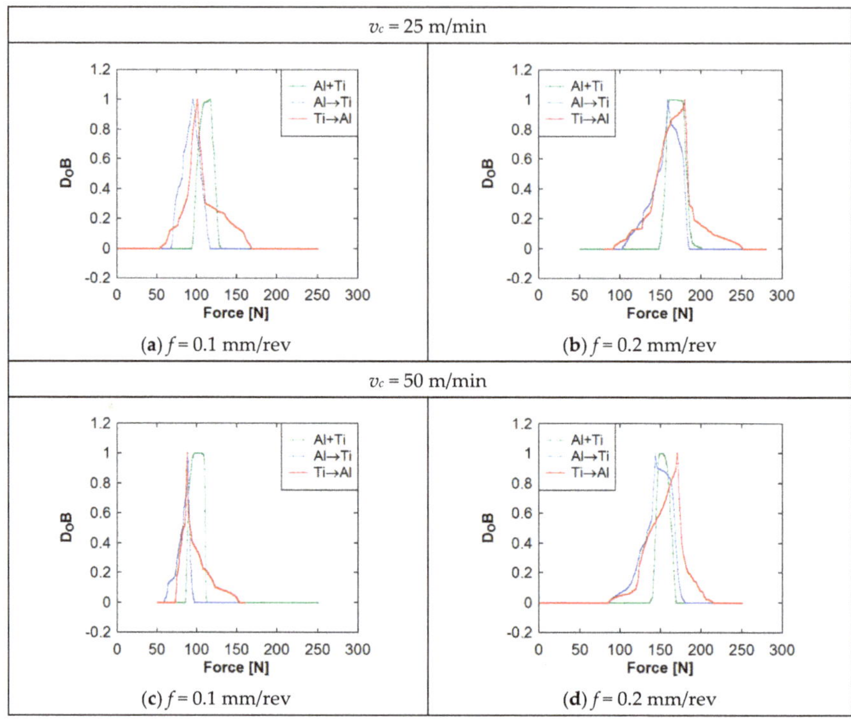

Figure 7. Uncertainties in the machining forces underlying Al–Ti.

5. Analyzing Machining Forces Underlying Al–CI

This section describes the machining forces underlying the bimetallic specimens denoted as Al–CI. It is worth mentioning that this is a uniform combination similar to the previous two cases, because the tensile strength, hardness, and percent elongation of cast iron are greater than those of aluminum (A5052), as listed in Table 1 (note that the hardness equal to 79.2 HRB is about 142 HV.) Compared to the previous case, the Al alloy used here is much harder. As such, it will help validate the conclusions made in the previous two sections.

Figure 8 shows the machining forces (thrust, feed, and cutting forces) in the time domain for the dissimilar metallic specimens denoted as Al–CI for the cutting conditions listed in Table 1. The plots in Figure 8a,d,g,j show the expected machining forces of the constituent materials (Al + CI), neglecting the joint area. The plots in Figure 8b,e,h,k show the machining forces manifested in the joint area while machining from the Al direction to the CI direction. The plots in Figure 8c,f,i,l show the machining forces manifested in the joint area while machining from the CI direction to the Al direction. As seen in Figure 8, if a low feed (0.1 mm/rev) and low cutting speed (25 m/min) are used, both machining directions provide similar cutting forces. For the high cutting speed, the machining direction soft-to-hard provides a better result only for the low feed. To be more specific, the uncertainty in the cutting forces shown in Figure 8 was further studied by constructing the possibility distributions similar to the previous two cases. The results are shown in Figure 9. The possibility distributions also support the abovementioned conclusions regarding the relationships between cutting conditions and cutting forces. In particular, the possibility distributions show that the use of a low feed and low cutting speed and using the cutting direction soft-to-hard is the best procedure for reducing the cutting force and its uncertainty.

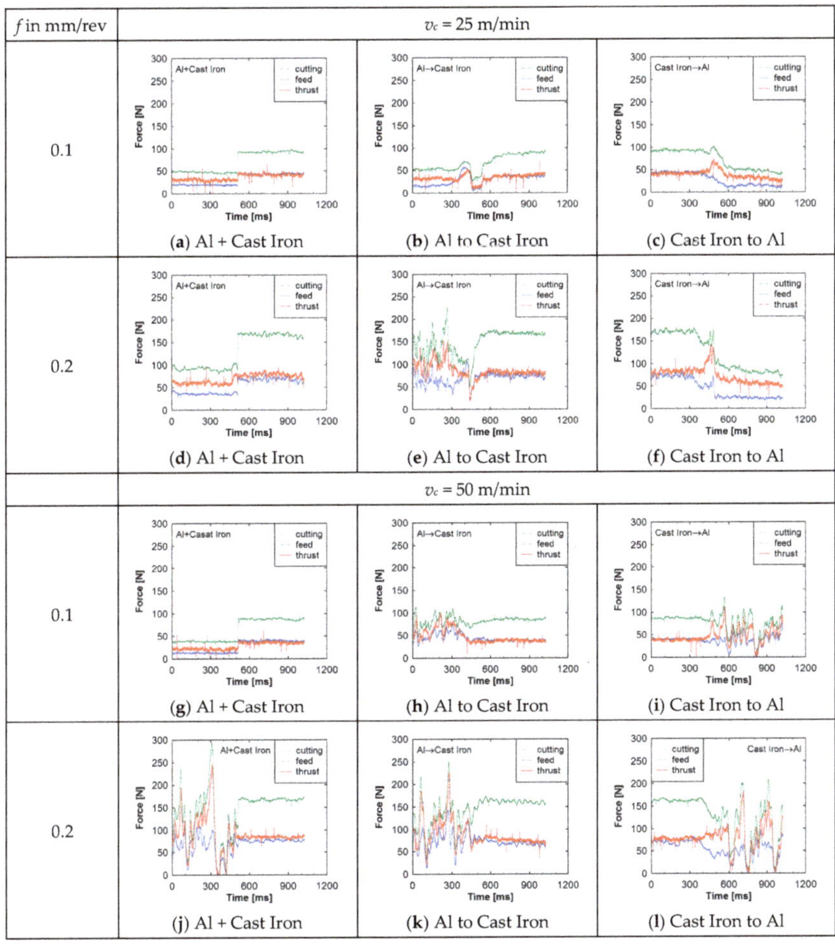

Figure 8. Machining forces underlying Al–CI.

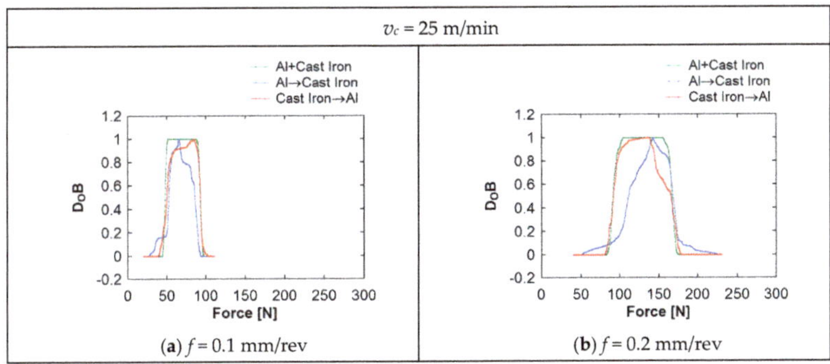

Figure 9. *Cont.*

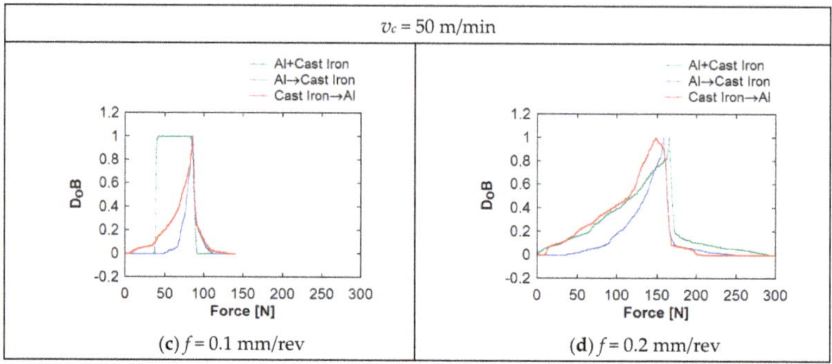

Figure 9. Uncertainties in the machining forces underlying Al–CI.

6. Discussions

Manufacturers who support sustainable product development (including design, manufacturing, and assembly) can benefit from the outcomes of this study because parts/products made of dissimilar materials (or multi-material objects) are better than their mono-material counterparts in terms of sustainability (cost, weight, and CO_2 footprint). Particularly, this kind of study will help them by supplying the knowledge of material wastages and energy conceptions during the manufacturing processes. Regarding the material wastage calculation, the methodology described in [2] can be used. As far as the energy consumption is concerned, the machining force signals shown in Figures 4–9 can be used. For example, the machining power (P_M) (kW) can be estimated using the cutting and feed force signals, which is a useful piece of information for determining the energy efficiency of a manufacturing process [2]. The machining power, denoted as P_M, has two components, namely, Cutting power (P_c) and Feed power (P_f) components. As such, the following formulation holds:

$$P_M(i) = P_c(i) + P_f(i) = \frac{1}{60 \times 10^3}\left[F_{SC}(i)v_c + \frac{F_{SF}(i)fN}{10^3}\right] \tag{1}$$

Figure 10 shows, for example, the P_M of the bimetallic specimen called SU–SC for the cutting conditions v_c = 25 m/min and f = 0.2 mm/rev. As seen in Figure 10, P_M varies in the range of [0.2, 0.45] kW. The variability in the cutting power for the four possibilities are illustrated in Figure 10a–d that correspond to the segments S15CK, SUS304, S15CK to SUS304, and SUS304 to S15CK, respectively. When the cutting tool passes the joint area, a gradual decrease/increase in the cutting power is observed, which is similar to that of the machining forces. This means that when the force sensors are not available, a power measurement instrument can be used to monitor the machining behavior of a bimetallic object.

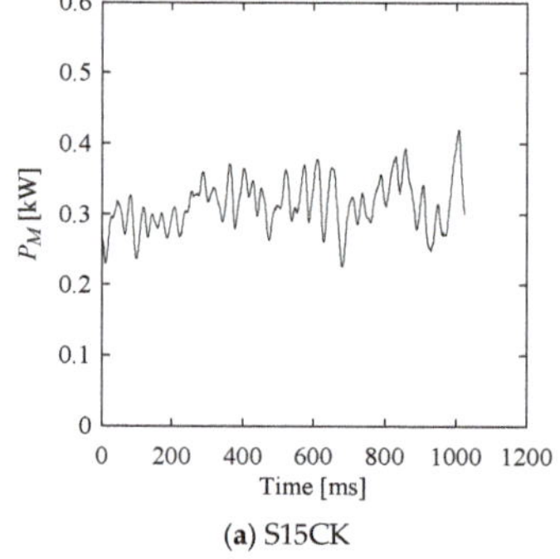

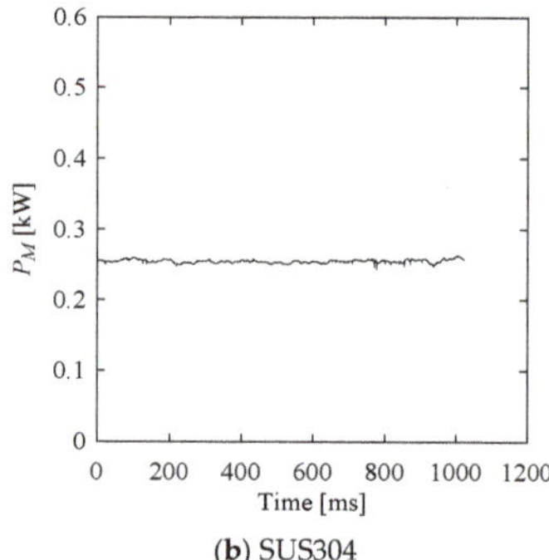

Figure 10. *Cont.*

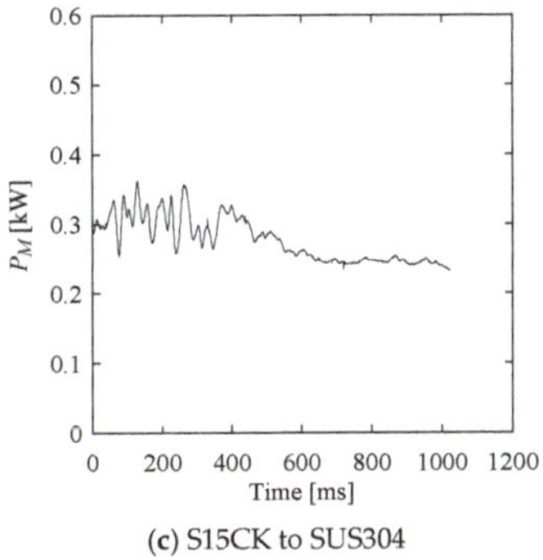
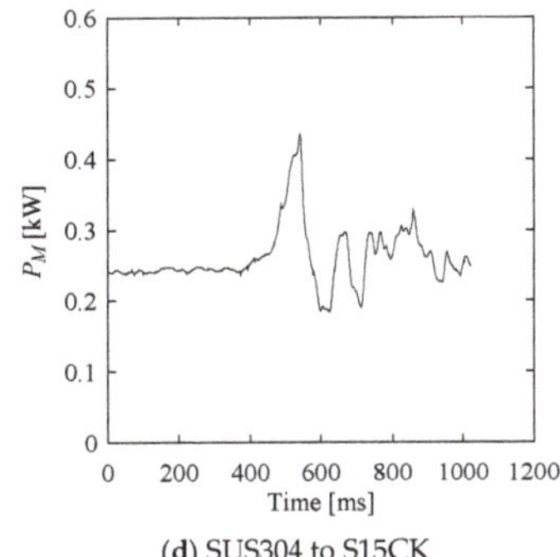

(c) S15CK to SUS304   (d) SUS304 to S15CK

Figure 10. Machining power of the SU–SC bimetallic specimen (v_c = 25 m/min, f = 0.2 mm/rev).

7. Concluding Remarks

This study reports the cutting/feed/thrust forces exhibited by three sets of bimetallic specimens. It was found that an entirely different machining force behavior arises due to the presence of two different materials, as well as the joint area.

The results shown in Figures 4–9 lead to the following conclusions:

Referring to the results in Figures 4 and 5, while machining steel-based bimetallic objects, keeping a low feed and high cutting speed is the better option, and the machining operation can be performed in both hard-to-soft and soft-to-hard material directions, but machining in the soft-to-hard material direction is the better option.

It is not recommended to create a bimetallic object using very soft material. Otherwise, it creates a machining problem (e.g., the case shown in Figures 6 and 7).

If an aluminum-based bimetallic part is preferred, then it is better to use a relatively harder alloy (e.g., compare the results shown in Figures 6 and 7 with those of shown in Figures 8 and 9). For the aluminum-based bimetallic objects, it is better to machine at a low cutting speed and low feed when the hard-to-soft material direction is needed.

Nevertheless, the research on machining is mostly concerned with the machining of objects made of mono-material and special alloys, whereas the research on machining objects made of multiple materials is in its infancy. The outcomes of this study can be used as a reference while enriching the machining technology of multi-material parts.

Funding: This work was funded by Kitami Institute of Technology.

Acknowledgments: The author gratefully acknowledges two of his former graduate students, Shin Matsui and Dongyuan Wu, and Masaaki Kimura at the University of Hyogo for their valuable inputs during the course of this study.

Conflicts of Interest: The author declares no conflict of interest.

Appendix A   Pictures of the Bimetallic Specimens Taken after Machining

This Appendix shows the pictures of the three types of specimens after conducting the turning experiments. The respective cutting conditions and directions are shown.

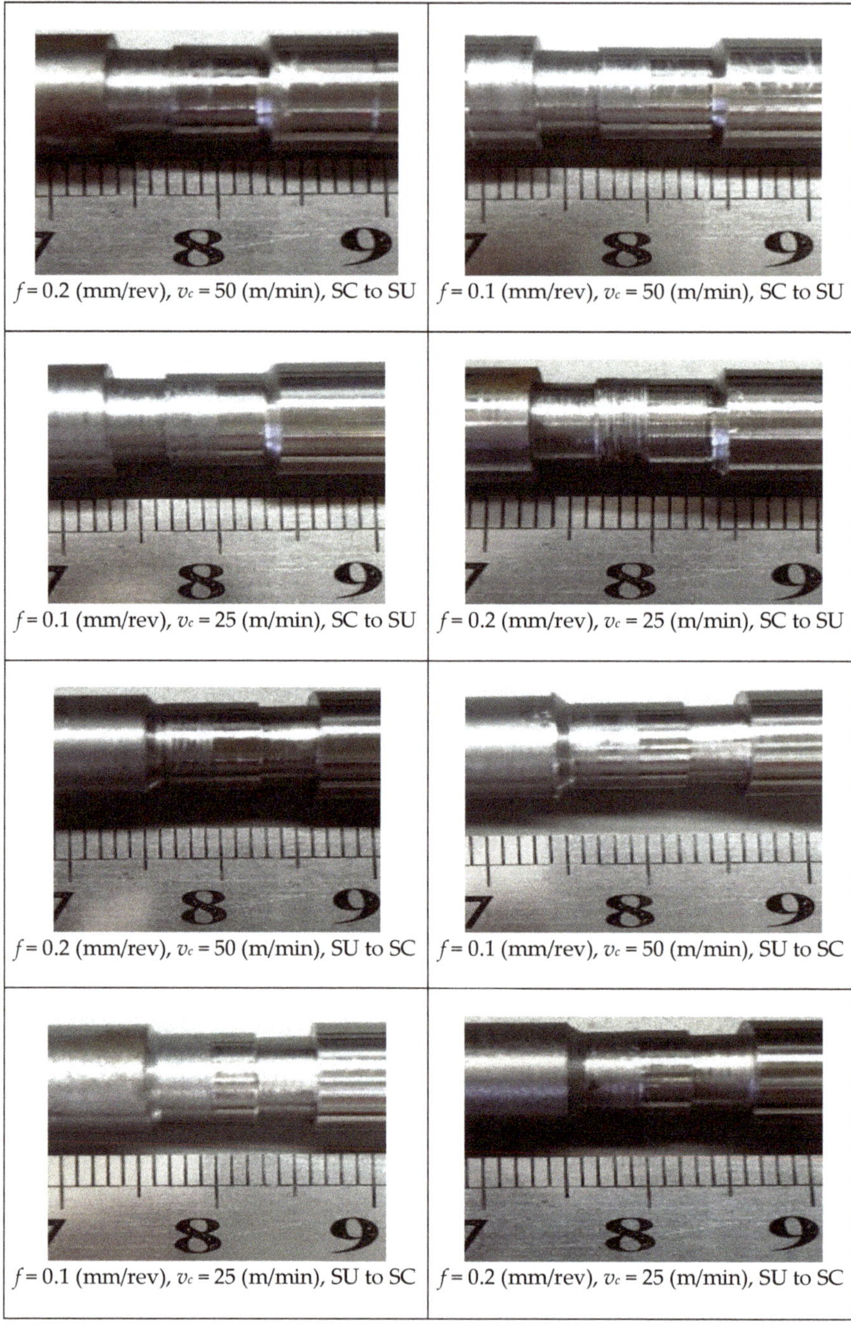

Figure A1. The SU–SC specimens.

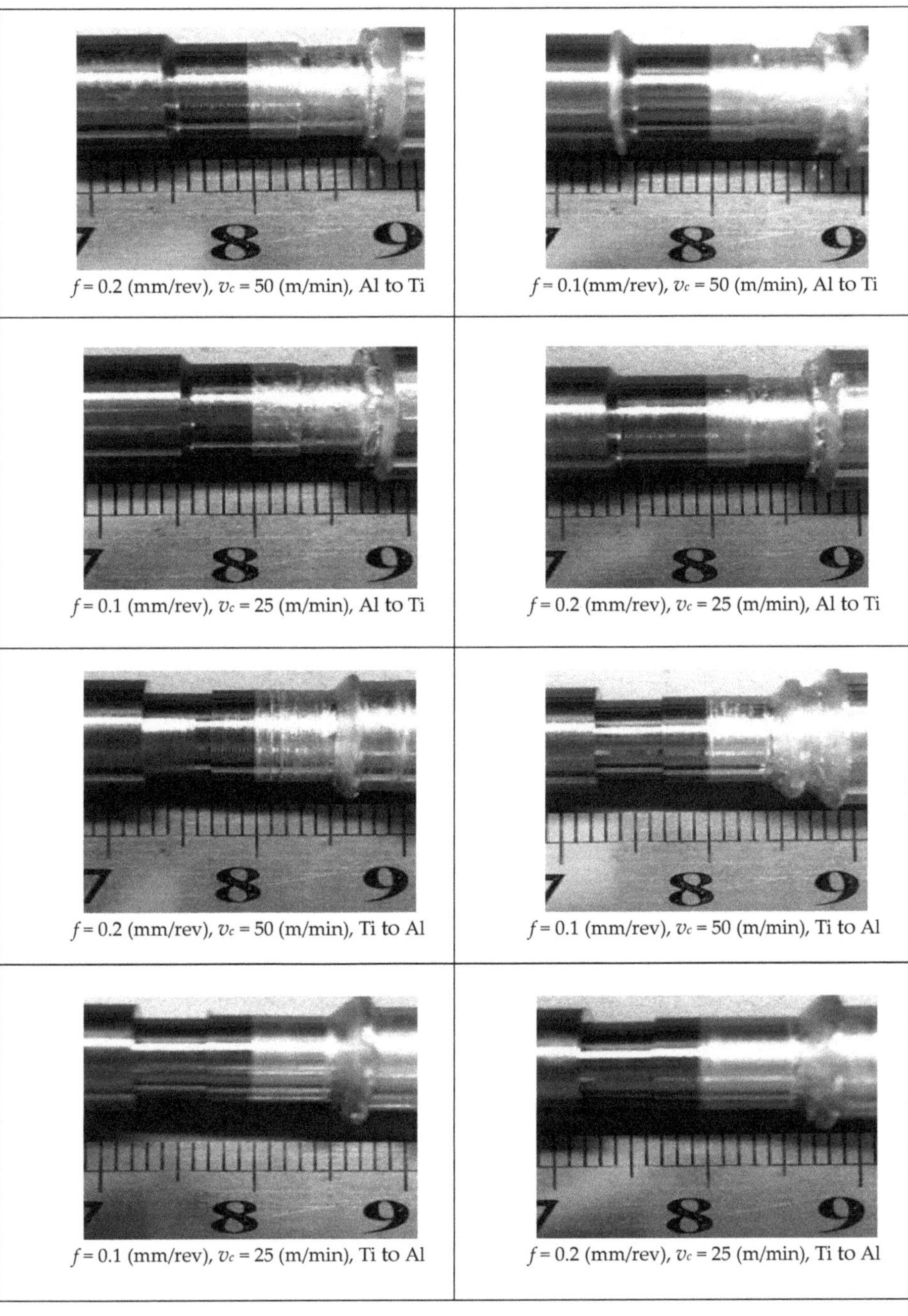

Figure A2. The Al–Ti specimens.

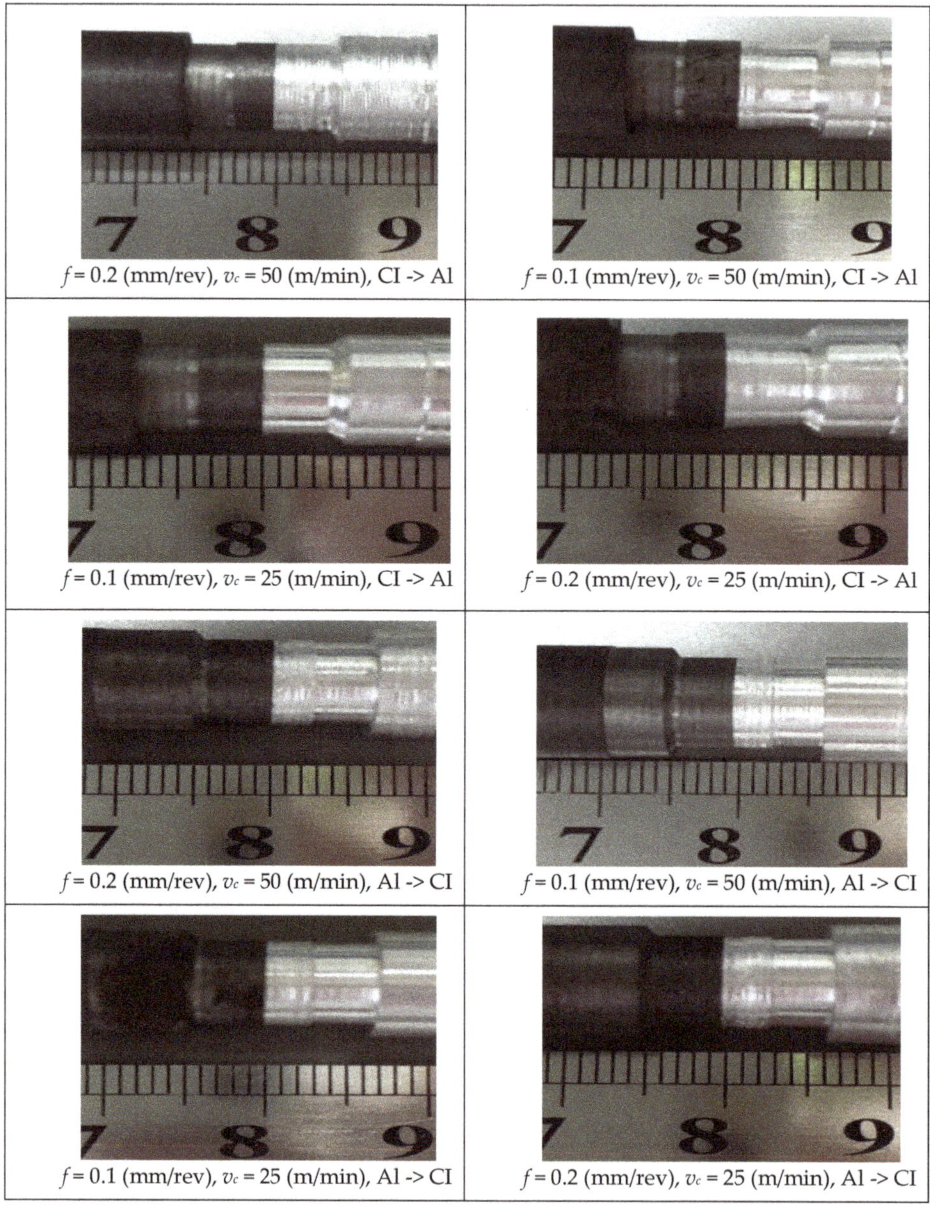

Figure A3. The Al–CI specimens.

Appendix B Inducing Possibility Distributions (Fuzzy Numbers) from Numerical Data

This appendix describes the mathematical procedures used to induce the possibility distributions (fuzzy numbers) from the time series of machining forces. The same procedure can be found in [25,26].

Let $x(t) \in \Re$, $t = 0, \dots, n - 1$ be n data points in the form of a time series, as shown in Figure A4.

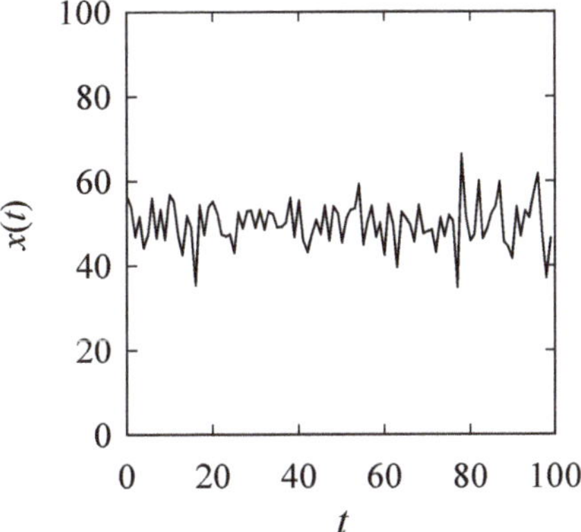

Figure A4. A given set of numerical data.

Let $(x(t), x(t+1))$, $t = 0, \ldots, n-1$ be a point-cloud in the universe of discourse $X = [x_{min}, x_{max}]$ so that $x_{min} < \min(x(t) \mid \forall t \in \{0, \ldots, n\})$ and $x_{max} > \max(x(t) \mid \forall t \in \{0, \ldots, n\})$. Let A and B two square boundaries so that the vectors of the vertices of A and B (in the anti-clockwise direction) are $((x_{min}, x_{min}), (x, x_{min}), (x, x), (x_{min}, x))$ and $((x_{max}, x_{max}), (x, x_{max}), (x, x), (x_{max}, x))$, respectively, $\forall x \in X$. As such, (x, x) is the common vertex of A and B. For example, consider the arbitrary point-cloud shown in Figure A5. According to Figure A5, the universe of discourse is as follows, $X = [20, 80]$. Notice the relative positions of the boxes denoted as A and B in Figure A5. The boxes are connected at their common vertices.

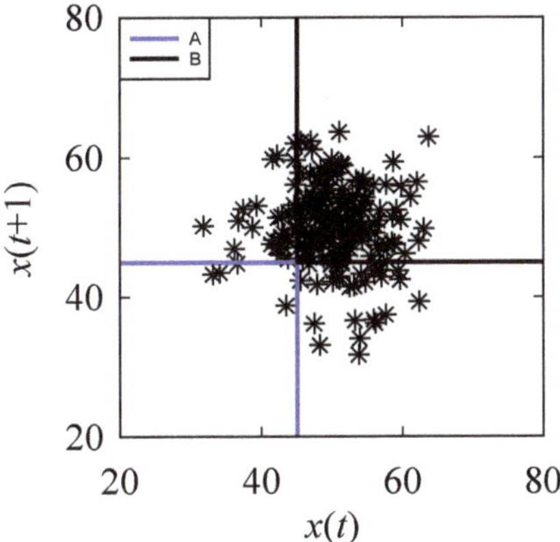

Figure A5. Relative position of A and B in the point-cloud $(x(t), x(t+1))$.

Let $Pr_A(x)$ and $Pr_B(x)$ be two subjective probabilities, wherein $Pr_A(x)$ and $Pr_B(x)$ represent the degrees of chance that the points in the point-cloud are in A and B, respectively. As such, these functions are defined by the following mappings:

$$X \to [0, 1]$$

$$x \mapsto Pr_A(x) = \frac{\sum\limits_{i=0}^{n-1} \Theta(t)}{n-1} \tag{A1}$$

$$\Theta(t) = \begin{cases} 1 & ((x(t) \leq x) \wedge (x(t+1) \leq x)) \\ 0 & otherwise \end{cases}$$

$$X \rightarrow [0,1]$$

$$x \mapsto \Pr_B(x) = \frac{\sum\limits_{i=0}^{n-1} \Omega(t)}{n-1}$$

$$\Omega(t) = \begin{cases} 1 & ((x(t) \geq x) \wedge (x(t+1) \geq x)) \\ 0 & otherwise \end{cases} \tag{A2}$$

The typical natures of the functions defined in Equations (A1) and (A2) are illustrated in Figure A6, using the information of the point-cloud shown in Figure A5. Note that $\Pr_A(x)$ increases with the increase in x, and the opposite is true for $\Pr_B(x)$. It is worth mentioning that $\Pr_A(x) + \Pr_B(x) \leq 1$ for the point-cloud, though for some cases, $\Pr_A(x) + \Pr_B(x) = 1$ (see Figure A7). This means that the expression $\Pr_A(x) + \Pr_B(x)$ does not serve the role of "cumulative probability distribution". A cumulative probability distribution can, however, be formulated by using the information of $\Pr_A(x)$ and $\Pr_B(x)$, as shown in Figure A7.

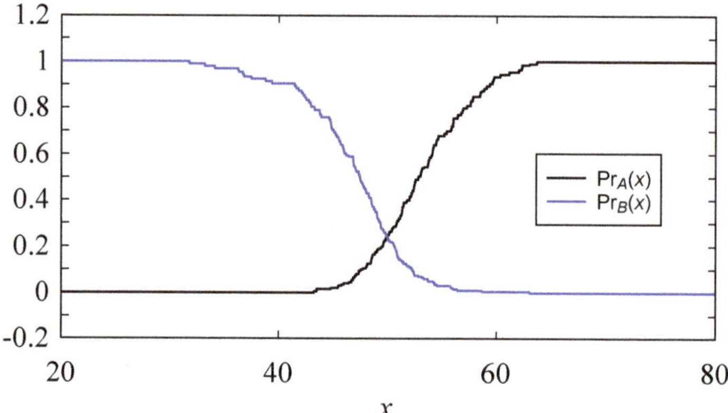

Figure A6. The typical nature of $\Pr_A(x)$ and $\Pr_B(x)$ for unimodal quantity.

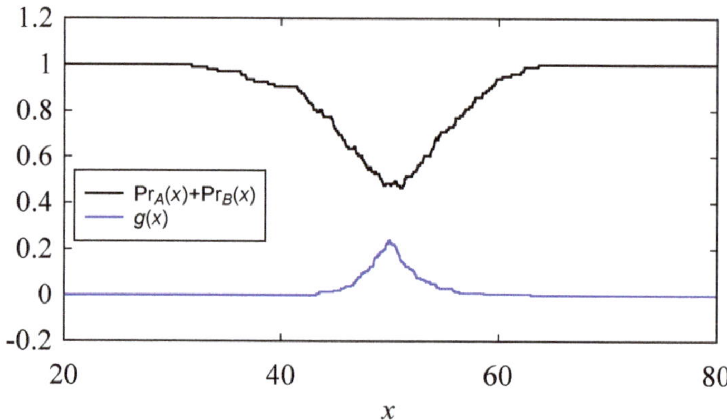

Figure A7. Nature of $\Pr_A(x) + \Pr_B(x)$ and $\min(\Pr_A(x), \Pr_B(x))$ for unimodal data.

Consider a mapping that maps x into the minimum of $\Pr_A(x)$ and $\Pr_B(x)$, as follows:

$$X \rightarrow [0,a]$$
$$x \mapsto g(x) = \min(\Pr_A(x), \Pr_B(x)) \tag{A3}$$

In Equation (A3), $a = 1$ if the point-cloud is a point; otherwise, $a < 1$. Figure A7 shows the nature of $g(x)$ with respect to $Pr_A(x) + Pr_B(x)$. The area under $g(x)$ is given by:

$$Q = \int_X g(x)dx \tag{A4}$$

There is no guarantee that $Q = 1$. Otherwise, $g(x)$ could have been considered a probability distribution of the underlying point-cloud. However, a function $F(x)$ can be defined as follows:

$$[0, a] \rightarrow [0, 1]$$
$$x \mapsto F(x) = \frac{\int_{x_{min}}^{x} g(x)dx}{Q} \tag{A5}$$

$F(x)$ can be considered a cumulative probability distribution because $\max(F(x)) = 1$, $F(x) \geq F(z)$ for $x \geq z$, $F(x) \in [0, 1]$, $\forall x, z \in X$. Figure A8 shows the nature of $F(x)$ derived from $g(x)$ shown in Figure A7. The cumulative probability distribution defined in Equation (A5) produces a probability distribution $Pr(x)$. Thus, the following formulation holds:

$$Pr(x) = \frac{dF(x)}{dx} \tag{A6}$$

Figure A9 shows the probability distribution $Pr(x)$ that corresponds to $F(x)$ as shown in Figure A8. The area under the probability distribution $Pr(x)$ is unit and $Pr(x)$ remains in the bound of $[0, 1]$.

From the induced probability distribution $Pr(x)$, a possibility distribution given by the membership function $\mu_I(x))$ can be defined based on the heuristic rule of probability-possibility transformation—the degree of possibility is greater than or equal to the degree of probability. The easiest formulation is to normalize $Pr(x)$ by its maximum value, $\max(Pr(x) \mid \forall x \in X)$, yielding the following formulation:

$$[0, 1] \rightarrow [0, 1]$$
$$Pr(x) \mapsto \mu_I(x) = \frac{Pr(x)}{\max(Pr(x)|\forall x \in X)} \tag{A7}$$

Figure A10 shows the possibility distribution $\mu_I(x)$ derived from the probability distribution $Pr(x)$ shown in Figure A9. The shape of the induced probability and possibility distributions are identical, as evident from Figures A9 and A10, respectively. Other formulations can be used instead of the formulation (A7), if needed.

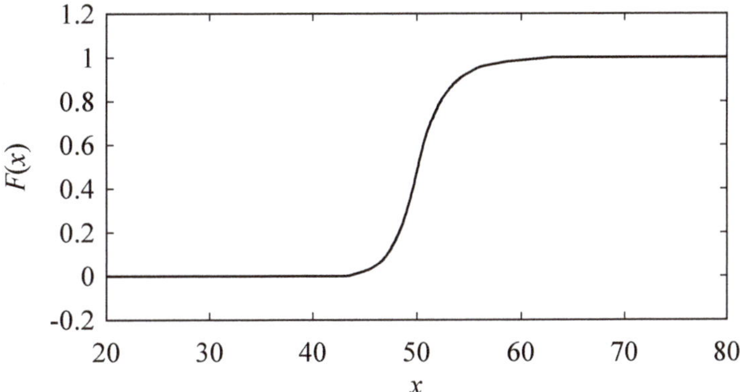

Figure A8. Nature of cumulative probability distribution of a point-cloud.

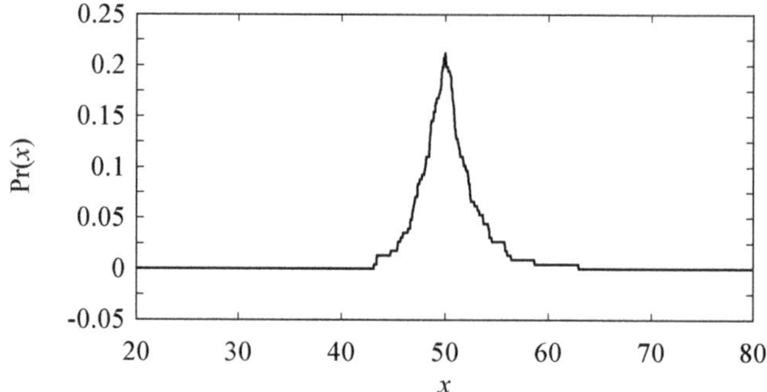

Figure A9. The nature of the probability distribution of a unimodal point-cloud.

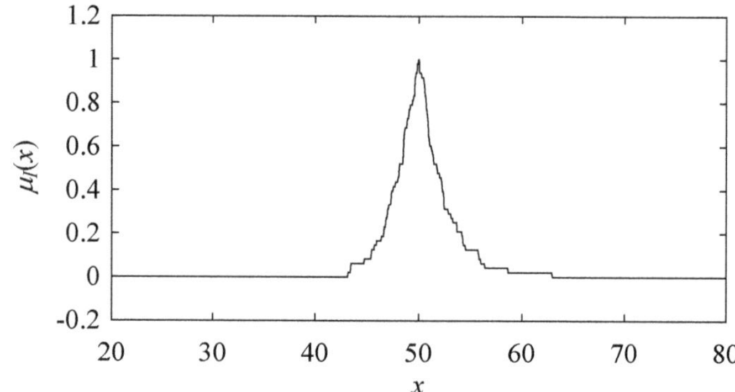

Figure A10. The nature of the possibility distribution of a unimodal point-cloud.

However, it is observed that when the point-cloud resembles the point-cloud of a bimodal quantity, the induced possibility distribution resembles a trapezoidal fuzzy number. In addition, when the point-cloud is a point, the induced possibility distribution becomes a fuzzy singleton. Moreover, when the point-cloud resembles the point-cloud of unimodal data, the induced probability/possibility distribution resembles a triangular fuzzy number. To define the membership function of an induced fuzzy number in the form of a triangular fuzzy number, the following formulation can be used.

Let u, v, and w be three points in ascending order in the universe of discourse X, $u \leq v \leq w \in X$. Let the interval $[u, w]$ be the *support* of a triangular fuzzy number and the point v be the *core*. The following procedure can be used to determine the values of u, v, and w from the induced fuzzy number $\mu_I(x)$ (Equation (A7)):

$$
\begin{aligned}
&u \leq v \leq w \in X \\
&u = x \quad (\mu_I(x) = 0 \wedge \mu_I(x + dx) > 0) \\
&v = x \quad (\mu_I(x - dx) < 1 \wedge \mu_I(x) = 1) \\
&w = x \quad (\mu_I(x - dx) > 0 \wedge \mu_I(x) = 0)
\end{aligned}
\tag{A8}
$$

As defined in (A8), u is the point after which the membership value $\mu_I(x)$ is greater than zero, v is the point corresponding to the maximum membership value $\max(\mu_I(x))$, and w is the point from/beyond which the membership value $\mu_I(x)$ again becomes/remains zero. Thus, the membership function of the induced triangular fuzzy number denoted as $\mu_T(x)$ is as follows:

$$
\begin{aligned}
&X \rightarrow [0, 1] \\
&x \mapsto \mu_T(x) = \max\left(0, \min\left(\tfrac{x-u}{v-u}, \tfrac{w-x}{w-v}\right)\right)
\end{aligned}
\tag{A9}
$$

In this article, the formulations up to (A7) were used, i.e., the regular fuzzy number was not constructed. The triangular fuzzy numbers are particularly important when the optimization of cutting conditions is carried out using the experimental data obtained by using a statistical procedure (e.g., design of experiment), as shown in [26].

References

1. United Nations Environment Program. *Report of the World Commission on Environment and Development Annex to General Assembly Document a/42/427*; United Nations: Nairobi, NY, USA, 1987.
2. Ullah, A.M.M.S.; Fuji, A.; Kubo, A.; Tamaki, J. Analyzing the sustainability of bimetallic components. *Int. J. Autom. Technol.* **2014**, *8*, 745–753. [CrossRef]
3. Allwood, J.M.; Ashby, M.F.; Gutowski, T.G.; Worrell, E. Material efficiency: A white paper. *Resour. Conserv. Recycl.* **2011**, *55*, 362–381. [CrossRef]
4. Bian, J.; Mohrbacher, H.; Zhang, J.-S.; Zhao, Y.-T.; Lu, H.-Z.; Dong, H. Application potential of high performance steels for weight reduction and efficiency increase in commercial vehicles. *Adv. Manuf.* **2015**, *3*, 27–36. [CrossRef]
5. Ullah, A.M.M.S.; Hashimoto, H.; Kubo, A.; Tamaki, J. Sustainability analysis of rapid prototyping: Material/resource and process perspectives. *Int. J. Sustain. Manuf.* **2013**, *3*, 20–36. [CrossRef]
6. Kimura, M.; Kusaka, M.; Kaizu, K.; Fuji, A. Effect of post-weld heat treatment on joint properties of friction welded joint between brass and low carbon steel. *Sci. Technol. Weld. Join.* **2010**, *15*, 590–596. [CrossRef]
7. Kimura, M.; Fuji, A.; Konno, Y.; Itoh, S.; Kim, Y.C. Investigation of fracture for friction welded joint between pure nickel and pure aluminium with post-weld heat treatment. *Mater. Des.* **2014**, *57*, 503–509. [CrossRef]
8. Sahu, P.K.; Kumari, K.; Pal, S.; Pal, S.K. Hybrid fuzzy-grey-Taguchi based multi weld quality optimization of Al/Cu dissimilar friction stir welded joints. *Adv. Manuf.* **2016**, *4*, 237–247. [CrossRef]
9. Buffa, G.; de Lisi, M.; Sciortino, E.; Fratini, L. Dissimilar titanium/aluminum friction stir welding lap joints by experiments and numerical simulation. *Adv. Manuf.* **2016**, *4*, 287–295. [CrossRef]
10. Gibson, I.; Rosen, D.; Stucker, B. *Additive Manufacturing Technologies: 3D Printing, Rapid Prototyping, and Direct Digital Manufacturing*, 2nd ed.; Springer: New York, NY, USA, 2015.
11. Bourell, D.L.; Rosen, D.W.; Leu, M.C. The Roadmap for Additive Manufacturing and Its Impact. *Print. Addit. Manuf.* **2016**, *1*, 231–238. [CrossRef]
12. Ullah, A.M.M.S. Design for additive manufacturing of porous structures using stochastic point-cloud: A pragmatic approach. *Comput. Aided Des. Appl.* **2018**, *15*, 138–146. [CrossRef]
13. Tateno, T. Anisotropic Stiffness Design for Mechanical Parts Fabricated by Multi-Material Additive Manufacturing. *Int. J. Autom. Technol.* **2016**, *10*, 231–238. [CrossRef]
14. Song, X.; Lieh, J.; Yen, D. Application of small-hole dry drilling in bimetal part. *J. Mater. Process. Technol.* **2007**, *186*, 304–310. [CrossRef]
15. Uthayakumar, M.; Prabhaharan, G.; Aravindan, S.; Sivaprasad, J.V. Machining studies on bimetallic pistons with CBN tool using the Taguchi method—Technical communication. *Mach. Sci. Technol.* **2008**, *12*, 249–255. [CrossRef]
16. Uthayakumar, M.; Prabhakaran, G.; Aravindan, S.; Sivaprasad, J.V. Influence of Cutting Force on Bimetallic Piston Machining by a Cubic Boron Nitride (CBN) Tool. *Mater. Manuf. Process.* **2012**, *27*, 1078–1083. [CrossRef]
17. Manikandan, G.; Uthayakumar, M.; Aravindan, S. Machining and simulation studies of bimetallic pistons. *Int. J. Adv. Manuf. Technol.* **2013**, *66*, 711–720. [CrossRef]
18. Malakizadi, I.; Sadik, L. Nyborg, Wear Mechanism of CBN Inserts During Machining of Bimetal Aluminum-grey Cast Iron Engine Block. *Procedia CIRP* **2013**, *8*, 188–193. [CrossRef]
19. Ullah, A.M.M.S.; Fuji, A.; Kubo, A.; Tamaki, J.; Kimura, M. On the Surface Metrology of Bimetallic Components. *Mach. Sci. Technol.* **2015**, *19*, 339–359. [CrossRef]
20. Matsui, S.; Ullah, S.; Kubo, A.; Fuji, A. Cutting force signal processing for machining bimetallic components. In Proceedings of the International Conference on Leading Edge Manufacturing in 21st Century: LEM21, Tokyo, Japan, 18–22 October 2015. [CrossRef]
21. Wu, D.; Ullah, S.; Kubo, A.; Fuji, A. On the complexity in roughness quantification across bimetallic boundary. In Proceedings of the International Conference on Leading Edge Manufacturing in 21st Century: LEM21, Tokyo, Japan, 18–22 October 2015. [CrossRef]

22. Kaynak, Y.; Kitay, O. Porosity, Surface Quality, Microhardness and Microstructure of Selective Laser Melted 316L Stainless Steel Resulting from Finish Machining. *J. Manuf. Mater. Process.* **2018**, *2*, 36. [CrossRef]
23. Cabrera, C.G.; Araujo, A.C.; Castello, D.A. On the wavelet analysis of cutting forces for chatter identification in milling. *Adv. Manuf.* **2017**, *5*, 130–142. [CrossRef]
24. Ullah, A.M.M.S.; Akamatsu, T.; Furuno, M.; Chowdhury, M.A.K.; Kubo, A. Strategies for Developing Milling Tools from the Viewpoint of Sustainable Manufacturing. *Int. J. Autom. Technol.* **2016**, *10*, 727–736. [CrossRef]
25. Ullah, A.M.M.S.; Shamsuzzaman, M. Fuzzy Monte Carlo Simulation using point-cloud-based probability–possibility transformation. *Simulation* **2013**, *89*, 860–875. [CrossRef]
26. Chowdhury, M.A.K.; Sharif Ullah, A.M.M.; Anwar, S. Drilling High Precision Holes in Ti6Al4V Using Rotary Ultrasonic Machining and Uncertainties Underlying Cutting Force, Tool Wear, and Production Inaccuracies. *Materials* **2017**, *10*, 1069. [CrossRef] [PubMed]

Journal of
Manufacturing and
Materials Processing

Article

Temperature- and Time-Dependent Mechanical Behavior of Post-Treated IN625 Alloy Processed by Laser Powder Bed Fusion

Alena Kreitcberg [1], Karine Inaekyan [1], Sylvain Turenne [2] and Vladimir Brailovski [1,*]

[1]   Department of Mechanical Engineering, École de Technologie Supérieure, 1100 Notre-Dame Street West, Montreal, QC H3C 1K3, Canada

[2]   Department of Mechanical Engineering, École Polytechnique de Montréal, 2900 boul. Édouard-Montpetit, Montreal, QC H3T 1J4, Canada

*   Correspondence: vladimir.brailovski@etsmtl.ca; Tel.: +1-(514)-396-8594; Fax: +1-(514)-396-8530

Received: 21 June 2019; Accepted: 26 August 2019; Published: 29 August 2019

Abstract: The microstructure and mechanical properties of IN625 alloy processed by laser powder bed fusion (LPBF) and then subjected to stress relief annealing, high temperature solution treatment, and hot isostatic pressing were studied. Tensile testing to failure was carried out in the 25–871 °C temperature range. Creep testing was conducted at 760 °C under 0.5–0.9 yield stress conditions. The results of the present study provided valuable insights into the static and creep properties of LPBF IN625 alloy, as compared to a wrought annealed alloy of similar composition. It was shown that at temperatures below 538 °C, the mechanical resistance and elongation to failure of the LPBF alloy were similar to those of its wrought counterpart, whereas at higher temperatures, the elongation to failure of the LPBF alloy became significantly lower than that of the wrought alloy. The solution-treated LPBF alloy exhibited significantly improved creep properties at 760 °C as compared to the wrought annealed alloy, especially under intermediate and low levels of stress.

Keywords: nickel-based superalloys; additive manufacturing; high temperature mechanical properties; creep resistance

---

1. Introduction

Nickel-based Inconel 625 alloy has numerous applications in the aeronautics, aerospace, marine, chemical, and petrochemical industries [1,2]. The alloy is generally used in a medium temperature range (250–593 °C) for structural applications requiring high strength and excellent corrosion resistance, and in a high temperature range (over 593 °C) for applications calling for outstanding creep resistance (ASTM E139-11, ASTM B444). These service properties can be achieved by conventional manufacturing technologies such as forging, rolling, or extrusion [3–6]; laser powder bed fusion (LPBF) additive manufacturing (AM) technology, however, offers numerous advantages over conventional manufacturing, more specifically in terms of its ability to fabricate parts with near net shapes, unique designs, added functionalities, low buy-to-fly ratios, and high productivity [7–10]. Moreover, LPBF is capable of producing fully functional parts directly from metal powder without the need for specialized tooling and intermediate processing steps.

It should nevertheless be noted that complex heat effects, which occur during LPBF and are related to highly localized multiple melting–remelting of powder and of underlying bulk materials, differ from those seen during conventional casting and welding. They are also responsible for strongly non-equilibrium heat and mass transfer and solidification phenomena leading to grain refinement, texture development, and the formation of unusual metastable phases [11–13]. High residual stresses resulting from a combination of significant temperature gradients and high cooling rates represent another peculiarity of LPBF; to avoid distortions and cracking, printed parts must therefore be subjected to post-processing stress relief heat treatment before they can even be removed from the build plate [14,15].

To mitigate the undesirable effects of LPBF processing on the microstructure of parts (columnar structure and precipitation formation), various post-processing heat treatments have been proposed [12,13,16,17]. These treatments frequently differ from those recommended for conventionally processed IN625 alloy parts, because of the previously mentioned structural features related to LPBF processing. For example, only a small amount of recrystallized structure is found in the LPBF IN625 alloy at 980 °C (~0.8 T_m) [12], while in the conventionally deformed ($\varepsilon = 0.4$) IN625 alloy, annealing in the 900–980 °C temperature range results in full recrystallization [18]. A fully recrystallized structure has been observed in the LPBF IN625 alloy only at temperatures higher than 1100 °C [12,13,16]. Thus, numerous studies on LPBF-fabricated alloys have aimed to find an original sequence of post-processing heat treatments, which can include solution treatment, homogenization annealing, aging, etc., in order to render the service properties of LPBF parts comparable or superior to those of conventionally manufactured alloys of similar compositions.

An excellent combination of outstanding corrosion resistance and superior creep resistance, as well as the relatively high tensile strength of nickel-based IN625 alloy (up to 600 °C), make it an interesting choice for aerospace applications. It has been shown that post-processing annealing of LPBF IN625 alloy can significantly improve its room temperature ductility as compared to its as-built state [13,19–21]. However, the assessment of mechanical properties cannot solely be limited to room temperature testing, especially for materials dedicated for service at elevated temperatures. In this context, it is known that conventionally processed nickel-based superalloys face the risk of embrittlement at temperatures higher than 600 °C, and that thermal treatments can affect this mechanical behavior either positively or negatively [22].

It has been shown, for example, that at 538 °C, the mechanical resistance and the elongation to failure of an IN625 alloy that was electron beam-melted and then hot isostatically pressed (HIP, 1120 °C, 100 MPa, 4 h) were close to those of wrought IN625 alloy [23]. At 760 °C, however, as compared to its wrought counterpart, the laser powder-fused IN625 alloy (HIP under the same conditions as above) manifested significantly lower ductility, but similar mechanical resistance [16,24]. Notwithstanding the preceding, such information is very limited, which makes it difficult to compare the tensile properties of printed and wrought IN625 parts. The outstanding creep resistance of wrought IN625 alloy favors its use at elevated temperatures, but, as was the case with the tensile properties, we could not find any publicly available information on the creep properties of 3D-printed IN625 alloy.

Unlike IN625 alloy, LPBF IN718 alloy, as a precipitation-hardened alloy with a higher mechanical resistance at elevated temperatures [25], has been covered by many studies [26–30]. It was shown that LPBF IN718 alloy manifested a high build-orientation-related anisotropy of its creep properties, caused by preferentially oriented distributions of dendrites and precipitations formed during LPBF processing [26]. Furthermore, the application of the solution (980 °C, 1 h) and aging (718 °C/8 h + 621 °C/10 h) heat treatments recommended by the AMS5662 specifications for forged and welded IN6718 alloy to the LPBF IN718 alloy led to lower creep rupture times, compared to the as-built state [26]. It was shown that this property degradation stems from the replacement of particle-shaped δ phase precipitates located in the interdendritic regions of the as-built alloy by needle-shaped δ phase precipitates located in the equiaxed structure of the solution-treated and aged alloy.

For the same LPBF IN718 alloy, limited data are available on its microstructure and mechanical behavior after post-processing HIP treatments. Similarly to the above-mentioned influence of the δ phase morphology, it was shown in Reference [28] that at 650 °C, the HIP-treated alloy (1200 °C, 103 MPa, 4 h) exhibited a lower creep rupture time compared to the as-built state. Note that this comparison was flawed, since the materials in both states were tested under the same stress of 650 MPa (~0.8 of YS for the as-built alloy), which put the HIP material under less favorable conditions, since HIP reduces the mechanical resistance characteristics of the printed material. Nevertheless, it was assumed that the needle-like δ phase grain boundary precipitates found in crept HIP-treated samples could be a cause of lower creep lifetimes, but their origins were not clear; it was uncertain whether the precipitation took place during HIP or if it occurred during creep testing at 650 °C. A comparison of the HIP LPBF IN718 alloy and the conventional hot-rolled IN718 alloy showed that the former manifested shorter creep lives than the latter under the same testing conditions [28].

It was also shown in Reference [27] that the application of the HIP conditions recommended for the wrought IN718 to the LPBF IN718 alloy (1180 °C, 175 MPa, 4 h [31]) triggered three concurrent phenomena, namely, microstructure homogenization, δ phase dissolution, and the formation of coarse carbide precipitates. When these phenomena were combined, they significantly improved the creep rupture time at 650 °C and 550 MPa, as compared to what was obtained in the as-built state.

Based on the above-mentioned observations, two main objectives can be established for future work: (a) Building a comprehensive database of the mechanical behaviors of LPBF IN625 alloy over a wide temperature range; for this study, the application of heat treatments recommended for the wrought alloy of the same composition was considered reasonable as a first approximation; (b) establishing a correlation between the mechanical properties of LPBF IN625 alloy and its microstructure (size of structural elements, nature and morphology of precipitates), with the ultimate goal of optimizing the post-processing conditions for this material.

This work focused on the first objective. The tensile and creep behaviors of laser powder-fused IN625 alloy subjected to stress relief (SR) annealing, solution treatment (ST), and hot isostatic pressing (HIP) were studied. The tensile behavior was studied in the 25 to 871 °C temperature range (68 to 1600 °F), with this range corresponding to the widest service diapason recommended for IN625 alloy [32]. The creep behavior was studied at 760 °C (1400 °F) under various stresses, this temperature corresponding to the onset of the high temperature embrittlement phenomenon observed in our previous work [16]. The correlation between the mechanical behavior and the structural features will form the core of the next publication.

It should be noted that since LPBF IN625 alloy in its as-built condition is characterized by a significantly high level of residual stresses and a strongly heterogeneous microstructure, it is not suitable for practical use, and was therefore excluded from consideration in this study. Meanwhile, the microstructure and the mechanical properties of the as-built LPBF IN625 alloy at room and elevated temperatures can be found elsewhere [16,24].

Note also that in this work, the exact heat and HIP treatment conditions have been omitted and the tensile testing stress values measured normalized to protect proprietary partner information.

2. Material and Methods

In this study, IN625 powder (EOS GmbH, Munich, Germany) with a chemical composition corresponding to UNS N06625 and ASTM B443 was used. An EOSINT M290 (EOS GmbH, Munich Germany) laser powder bed fusion system equipped with a 400 W ytterbium fiber laser and the EOS IN625_Surface 1.0 Parameter Set (laser power ~300 W, scanning speed ~1000 mm/s, hatching space ~0.1 mm, and layer thickness ~40 μm) was employed to fabricate two types of specimens: $10 \times 10 \times 10$ mm^3 cubic specimens for microstructure evaluation, and $85 \times 18 \times 3$ mm^3 rectangular blanks for tensile testing (Figure 1a). The blanks were built in two directions relative to the build plate, as defined in Figure 1a, and have been referred to as vertical or horizontal (parallel or perpendicular to the build direction, respectively) throughout this paper. The chemical compositions of the IN625 powder, the as-built LPBF alloy, and the wrought annealed alloy (reference) are shown in Table 1.

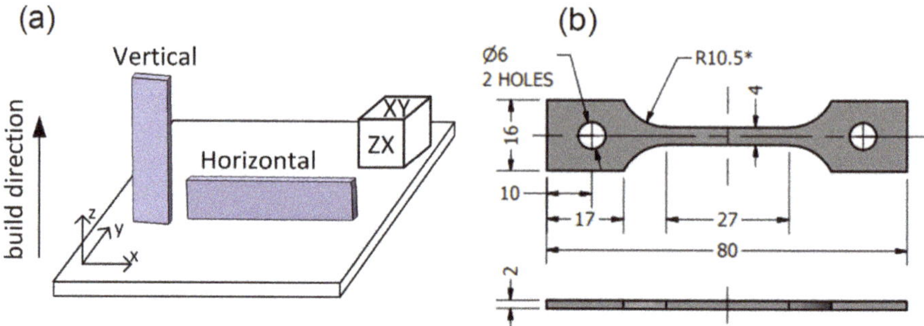

Figure 1. (**a**) Rectangular blanks and cubic specimens on the build plate; (**b**) tensile specimens (dimensions in mm; * - reference dimension) [24].

Table 1. Chemical composition of IN625 alloy (in wt. %).

IN625	Ni	Cr	Mo	Nb	Fe	Ti	Al	Co	C	Ta	Si	S	Mn
ASTM B443-00	Bal	21.0–23.0	8.0–10.0	3.15–4.15	≤5.0	≤0.4	≤0.4	≤1.0	≤0.1	≤0.05	≤0.5	≤0.015	-
Powder	Bal	21.81	9.33	4.06	0.78	0.39	0.34	0.19	0.013	<0.02	0.15	0.002	0.04
LPBF alloy	Bal	22.42	9.57	3.95	1.66	0.07	0.19	<0.03	0.012	0.02	0.17	0.004	0.04
Wrought alloy	Bal	23.67	8.49	3.49	4.48	0.22	0.24	0.07	0.032	0.02	0.18	0.001	0.34

Following the LPBF, the build plate with cubic and rectangular specimens was subjected to stress relief (SR) annealing at ~900 °C for 1 h (EOS recommendations), followed by forced air cooling (~1.5 °C/s) [10,33]. The SR treatments were carried out in a Nabertherm H41/N furnace under argon continuous flow (~15 L/min). Next, all the printed specimens were cut from the platform, using a reciprocated saw, and the rectangular blanks were machined by EDM (electrical discharge machining) to obtain the dumbbell-shaped tensile testing specimens shown in Figure 1b.

Finally, some SR specimens were reserved for future study, while the others were subjected to either hot isostatic pressing (HIP, Avure Technologies, Quintus QIH-3, Columbus, OH, USA) under pressurized argon atmosphere, followed by furnace cooling (~0.1 °C/s) [34–36]; or high temperature solution treatment (ST) for 1 h in an open-air furnace (Pyradia, Longueuil, QC, Canada), followed by air cooling (~0.5 °C/s) (Figure 2). Both the HIP and ST post-treatments were expected to homogenize the as-built LPBF microstructure and decrease the anisotropy of the IN625 alloy's mechanical properties. Since HIP is time- and resource-consuming, and can result in undesirable grain growth [16], ST is seen as its economic and technologically sound alternative.

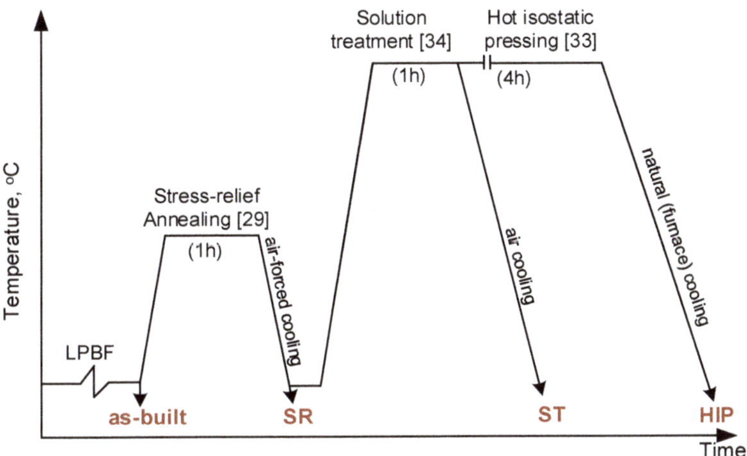

Figure 2. Schematic presentation of the post-processing sequence of laser powder bed fusion (LPBF) alloy.

For this study, it was decided to carry out both treatments at the same temperature in the 1100–1150 °C range. This decision can be explained by the fact that HIP at this temperature had already been successfully used to reduce processing-induced porosity, and to homogenize and recrystallize LPBF-built microstructures [16,24,34]. On the other hand, for a high carbon version of IN625 alloy (~0.045 wt. %C), a typical ST temperature is also in the 1100–1150 °C range [37].

For reference, wrought IN625 alloy annealed at 980 °C (Table 1), provided by McMaster Corp. and corresponding to ASTM B443 (Grade 1, with a grain size of ~13 µm), was also characterized in the framework of this study. The Grade 1 (fine-grained) alloy was chosen because it possessed higher mechanical characteristics above 600 °C than the solution-treated Grade 2 (coarse-grained) alloy. The temperature- and time-dependent behaviors of the LPBF and wrought IN625 alloys were compared in this study from the perspective of their concurrent industrial use.

Tensile testing with a strain rate of $10^{-3} \cdot s^{-1}$ was conducted at 25, 427, 538, 593, 649, 760, and 871 °C (68, 800, 1000, 1100, 1200, 1400, and 1600 °F) using an MTS 810 testing system equipped with an infrared radiant heating furnace. High temperature testing was realized under argon atmosphere at a flow rate of 5–18 L/h. Prior to tensile testing, specimens were heated at a heating rate of 1 °C/s and maintained at the test temperature for 10 min. The temperature was controlled using three K-type thermocouples in contact with the specimen surface and evenly distributed along its gauge length to control the uniformity of the temperature distribution. The strain was calculated using data provided by the LVDT (linear variable differential transducer) of the testing machine. After each treatment and for each testing temperature, the yield strength (YS corresponding to 0.2% offset strain), the ultimate tensile strength (UTS), and the elongation to failure (ε) were determined. For each experimental point, three specimens were tested, and the mean values of YS, UTS, and ε and their confidence ranges at a confidence probability of $p = 0.95$ were calculated.

Creep tensile testing was conducted at 760 °C (1400 °F) at 0.5, 0.7, and 0.9 of the YS with a loading rate of 10 N·s^{-1}. The testing system, atmosphere, gas flow, and heating rates for the creep testing were identical to those of the elevated temperature tensile testing. Three tests were conducted for each creep condition, and the rupture time (τ), the fracture strain (ε), and the steady or secondary creep rates ($\dot{\varepsilon}$) were determined.

The fracture morphology and microstructure were analyzed using scanning electron microscopy (SEM, Hitachi TM3030 system and Hitachi SU8230 system equipped with an electron backscatter diffraction (EBSD) unit). The microstructural analysis was performed on the horizontal (XY) and vertical (ZX) reference faces of the cubic specimens (Figure 1a). All the specimens were polished manually (down to 1 µm grit size), and then using a vibrometer and colloidal silica (0.05 µm grit size). For EBSD analysis, samples were tilted at 70° and scanned at 20 kV, with a step of 1–2 µm.

3. Results

3.1. Grain Size and Grain Orientation

The EBSD images of the SR-, ST-, and HIP-treated specimens are shown in Figure 3. In the vertical section, the SR specimen contained grains oriented parallel to the build direction (Figure 3a). A continuous grain growth (i.e., epitaxial growth across the melt pool boundaries) during the LPBF process affected the grain orientation and the length of growing grains. In fact, the grains had an average length greater than 120 µm (three layer thicknesses). Equiaxed grains with an average size of 20 µm were observed in the horizontal section. After the ST and HIP treatments, the columnar grain structure of the SR alloy morphed into equiaxed grain structures (Figure 3b,c). After the ST, the grain size varied from 1 to 80 µm in size, while after the HIP, it varied between 10 and 300 µm. Note that the average grain size (~45 µm) of the HIP alloy was twice as large as that of the ST alloy (~20 µm). The equiaxed grains with annealing twins corresponded to a low-stacking fault-free energy fcc matrix. For the SR specimens, the dominant grain texture in the build direction corresponded to the [001] direction (red area in Figure 3a), which transformed to a random texture after the ST and HIP treatments (Figure 3b,c).

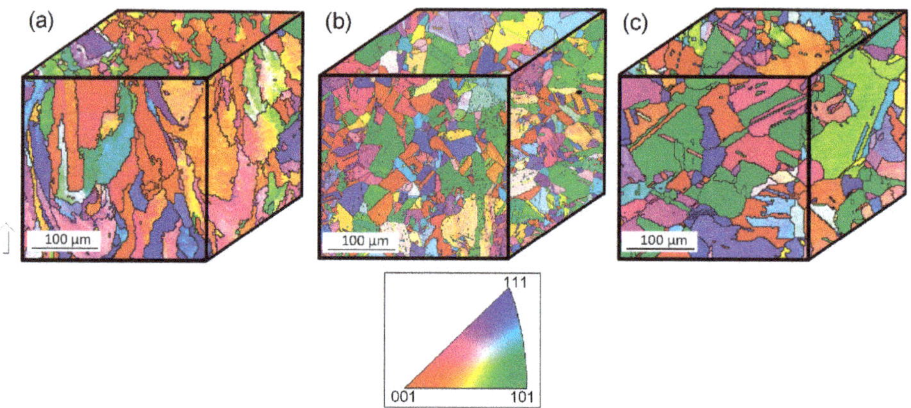

Figure 3. Electron backscatter diffraction (EBSD) images of the (**a**) stress-relief annealing (SR)-, (**b**) solution-treated (ST)-, and (**c**) hot isostatic pressing (HIP)-treated alloys; white arrow shows the build direction. (Color crystal orientation code is inserted.)

3.2. Mechanical Properties at Room Temperature

Typical RT tensile stress–strain diagrams of the SR-, ST-, and HIP-treated specimens are shown in Figure 4a–c, along with the stress–strain diagram of the wrought annealed alloy in Figure 4d. For the build orientation dependency evaluation, tensile diagrams for the horizontal and vertical specimens have been superimposed.

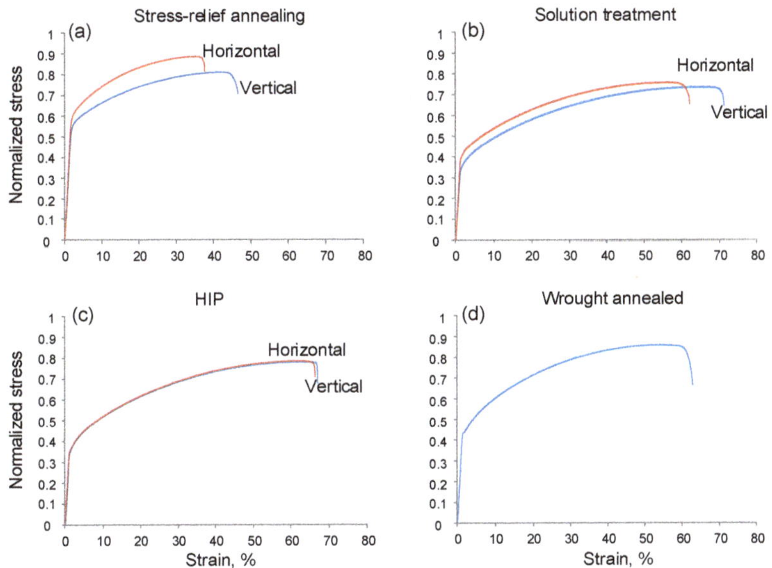

Figure 4. Tensile stress–strain diagrams (RT) of the (**a**) SR-, (**b**) ST-, and (**c**) HIP-treated LPBF alloy and (**d**) wrought annealed alloy.

An examination of these diagrams reveals that at room temperature, the SR specimens manifested the strongest build orientation dependency and the most hardened mechanical behavior as compared to their ST and HIP counterparts: after SR, the yield strength (YS) and the ultimate tensile strength (UTS) of the horizontal specimens were higher, while their elongations to failure (ε) were lower than those of the vertical specimens (Figure 4a). After ST, both the strength characteristics and their orientation dependency decreased (Figure 4b), while the elongations to failure increased, as compared to the corresponding values for the SR specimens. The YS and UTS values of the horizontal ST specimens were still slightly greater than those of the vertical ST specimens, while the elongations were lower. The HIP specimens did not manifest any orientation dependency and their elongations to failure were similar to those of the ST specimens, while their strength characteristics were slightly lower (Figure 4c).

3.3. Mechanical Properties in the 25–871 °C (1600 °F) Temperature Range

Typical stress–strain diagrams of the SR-, ST-, and HIP-treated LPBF specimens and the wrought annealed alloy are presented in Figure 5 for the temperature ranging from 25 to 871 °C (1600 °F).

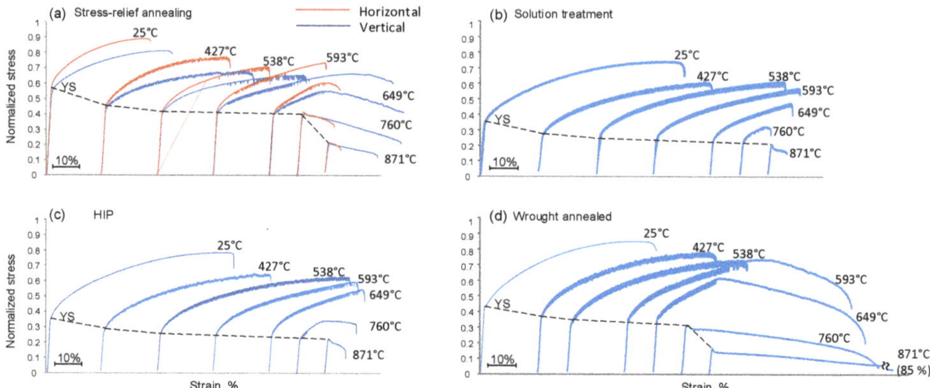

Figure 5. Stress–strain diagrams of the (**a**) SR-, (**b**) ST-, and (**c**) HIP-treated specimens and (**d**) wrought annealed alloy in the 25–871 °C temperature range.

Globally, at elevated temperatures, the horizontal SR specimens exhibited higher YS and UTS and lower elongations as compared to their vertical counterparts (Figure 5a), which was similar to the room temperature behavior of these specimens. It was seen, however, that the higher the testing temperature, the lower the orientation dependency in terms of the YS and UTS values, but the higher this dependency in terms of their elongations to failure. As shown in Reference [16], after low temperature ST and HIP treatments, the mechanical behaviors of the corresponding specimens became build orientation-independent, and therefore, only vertical specimens were tested in the present work (Figure 5b,c).

In the 427–538 °C temperature range, the SR specimens manifested the highest YS and UTS values as compared to those of the ST (HIP) and wrought alloy specimens, but at the expense of lower elongations. In this specific temperature region, the LPBF specimens (both ST and HIP) and wrought specimens manifested similar mechanical behavior. However, starting at 593 °C and up, the ductility of the ST and HIP specimens decreased, while that of the wrought alloy increased (the same trend was observed for the SR specimens at 649 °C and up). Vertical ST and HIP specimens manifested lower elongations than their vertical SR counterparts, but higher elongations than their horizontal SR counterparts. Thus, in the 593–649 °C temperature range, the mechanical behavior of the LPBF alloy was characterized by increasing brittleness, whereas the wrought alloy showed only a slight decrease in ductility.

At 760 °C, both the SR and wrought specimens exhibited yield strength peaks, while the ST and HIP specimens both manifested significant work hardening. At 871 °C, yield strength peaks and work softening were observed for all the specimens. At this temperature, the mechanical strength parameters of all the specimens were similar, while the elongations to failure of the LPBF alloy specimens were significantly lower than those of the wrought alloy, irrespective of the post-processing conditions of this study.

Additionally, the fracture analysis of the SR and HIP specimens after their tensile testing at 538 and 760 °C confirmed that in this temperature range, the specimens manifested fracture mechanisms corresponding to the transition from ductile to brittle behavior [24]. As seen in Figure 6, at 538 °C, the fracture surfaces showed a mostly ductile fracture mode characterized by dimples (Figure 6a,c), while at 760 °C, intergranular brittle fracture occurred (Figure 6b,d). Moreover, at 760 °C, after HIP, the fracture surface reflected visible triple points between equiaxed grains.

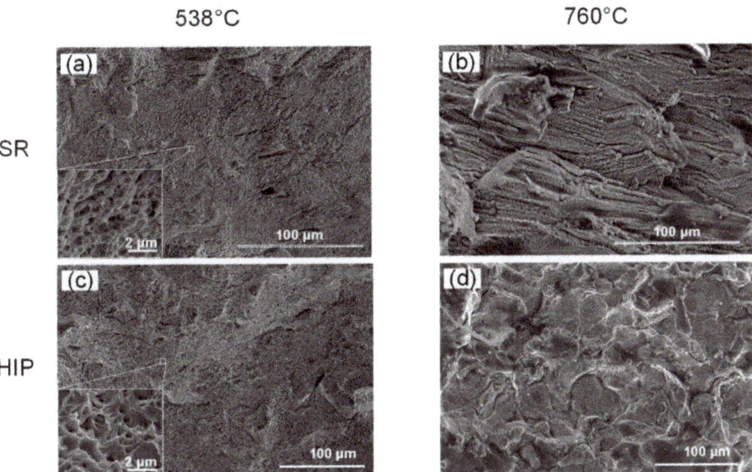

Figure 6. Fracture surfaces of the (**a,b**) SR (horizontal) and (**c,d**) HIP specimens, tensile-tested at (**a,c**) 538 °C and (**b,d**) 760 °C.

3.4. Creep Testing

Since the tensile mechanical properties of the ST and HIP specimens were found to be comparable across the entire temperature range of this study, only the ST-treated specimens were subjected to creep testing in accordance with ASTM E139–11 (2018), along with their SR and wrought annealed counterparts. Moreover, since the mechanical properties of the ST specimens over this temperature range were found to be build orientation-independent, only vertical ST specimens were subjected to this testing.

Figure 7 shows the typical creep diagrams for the SR, ST, and wrought annealed alloy specimens under stresses ranging from 0.5 to 0.9 YS (760 °C, 1400 °F). For the LPBF specimens subjected to low (0.5 YS) and intermediate (0.7 YS) stresses, three stages of creep behavior were clearly distinguished: primary, secondary (steady), and tertiary (fracture), with durations dependent on the levels of stress applied. In the case of the SR specimens (Figure 7a), a secondary stage of creep was less pronounced than in the case of the ST specimens (Figure 7b), and for the SR specimens at higher stresses (0.7 YS), the transition from the primary to the steady stage was hardly distinguishable. For the wrought annealed specimens, the primary stage of creep was almost skipped (Figure 7c); the steady stage started much more quickly: the start time was ≤ 0.003 h as compared to the SR and ST specimens, for which this time was 0.15–0.2 h.

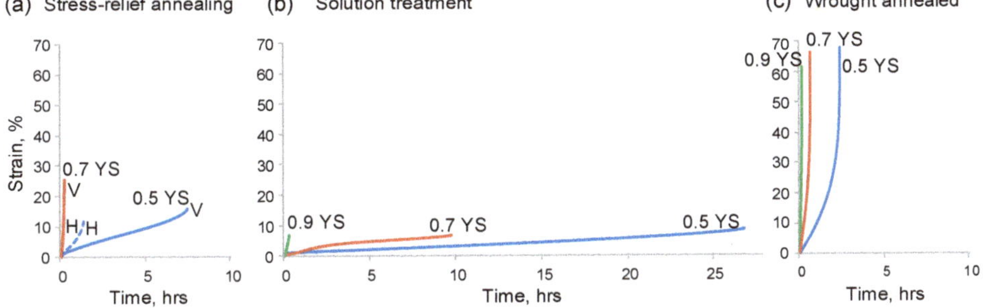

Figure 7. Creep diagrams of the (**a**) SR, (**b**) ST, and (**c**) wrought annealed IN265 alloys at 760 °C (1400 °F).

Figure 8 collects the lifetime and rupture strain data for the normalized levels of creep stress (σ/YS). Note large data scattering in the case of LPBF specimens, both SR and ST, as compared to their wrought counterparts (Figure 8a). In the 0.3–0.9 σ/YS range, the ST-treated LPBF alloy manifested a longer life to rupture (note also a run-out at 48 h at σ/YS < 0.5), while the SR and wrought specimens exhibited similar rupture times in the entire 0.3–0.9 σ/YS range. Interestingly, the mean rupture strain in the case of the wrought specimens was significantly higher than that in the case of the SR- and ST-treated LPBF specimens (Figure 8b). The ST specimens showed the lowest mean rupture strain. Note also that the mean rupture strain values correlated well with (and did not exceed) the corresponding tensile ductility values at 760 °C (Figure 5).

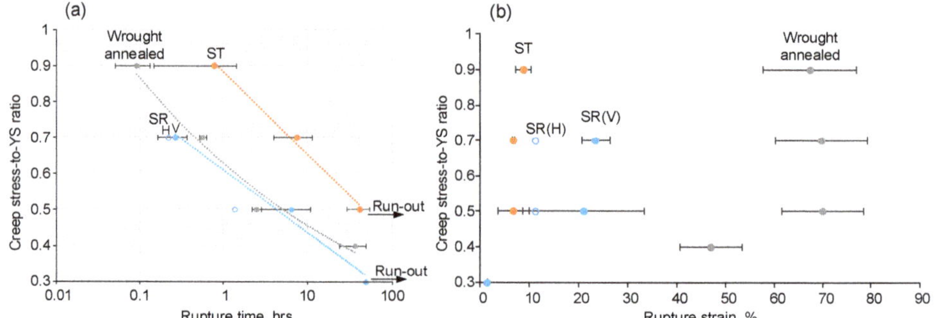

Figure 8. Applied stress in relative values (σ/YS) as a function of the (**a**) creep rupture time and (**b**) rupture strain for the SR, ST, and wrought annealed IN625 alloy (760 °C).

After low-stress creep fracture (0.5 YS) at 760 °C, the fracture surfaces of the SR specimens had a mix of transgranular and intergranular patterns, while the ST specimens contained mostly transgranular patterns corresponding to a dimpled fracture after the same stress creep (Figure 9a,c). High stresses, however, led to mostly intergranular fractures for both the SR and ST specimens (Figure 9b,d).

The fracture surfaces of the wrought annealed alloy, containing numerous voids, dimples, and tearing ridges, indicated extremely ductile behavior of this alloy as compared to that of the ST-treated LPBF alloy, under creep conditions at 760 °C (Figure 9e,f).

Finally, it is worth mentioning that the SR specimens exhibited a build orientation dependency of creep properties. The vertical specimens showed a more pronounced steady stage (Figure 7a) and a higher lifetime and rupture strain (Figure 8) as compared to their horizontal counterparts. Note also that the horizontal SR specimens demonstrated significantly lower tensile and creep properties than their vertical counterparts (Figures 5 and 7).

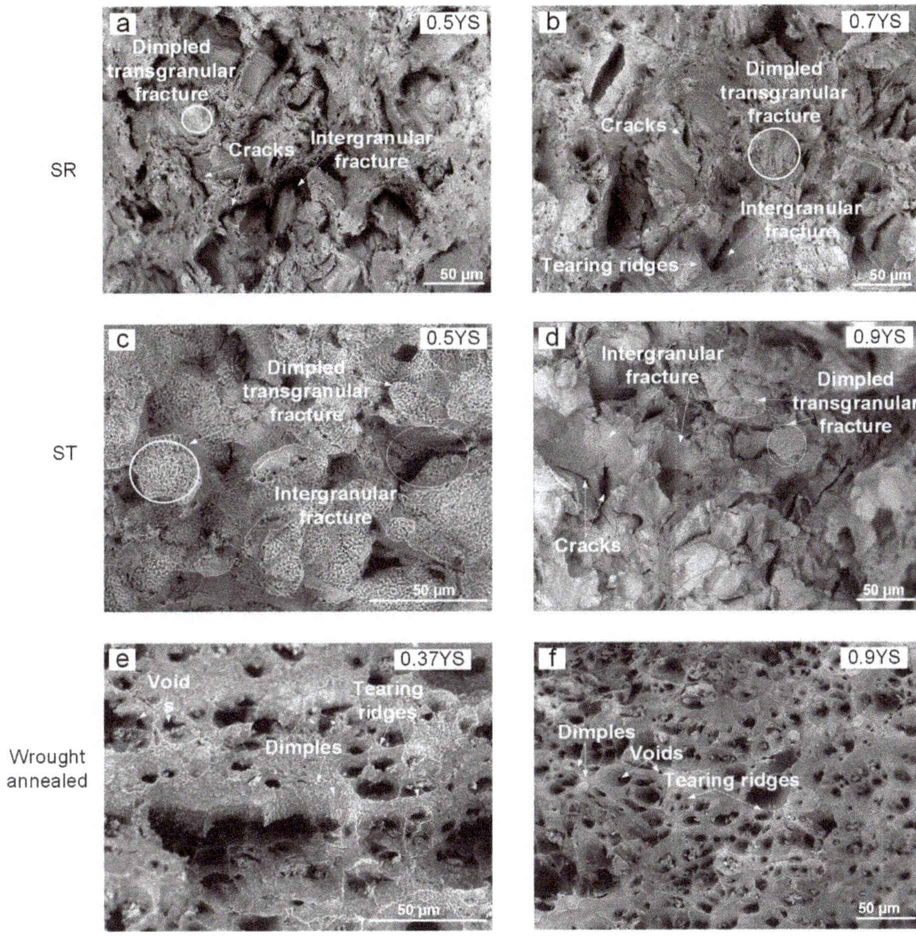

Figure 9. Scanning electron microscope (SEM) images of the fracture zones of creep specimens: (**a**,**b**) SR, (**c**,**d**) ST, and (**e**,**f**) wrought annealed IN625 alloys.

4. Discussion

Grain size and grain orientation. It was shown that LPBF processing was responsible for the formation of an anisotropic microstructure. The SR specimens contained grains elongated in the build direction with the dominant [001] texture, inherited from the as-built material. The grain length was more than twice the layer thickness, while the average width was about 20 µm. During the ST and HIP treatments, grain growth caused the microstructure to become nearly equiaxed with a random texture. The average grain sizes after the ST and HIP treatments were about 20 and 45 µm, respectively [16].

Map of the mechanical properties. The mechanical properties obtained at a $10^{-3} \cdot s^{-1}$ strain rate across a wide range of temperatures are shown in Figure 10a–c for the SR-, ST-, and HIP-treated LPBF IN625 alloys, respectively. The mechanical properties of the vertical and horizontal SR specimens have been presented here, while the properties of the ST and HIP specimens are limited to those of the vertical specimens. For reference, the mechanical properties of the wrought annealed alloy are also illustrated in Figure 10d.

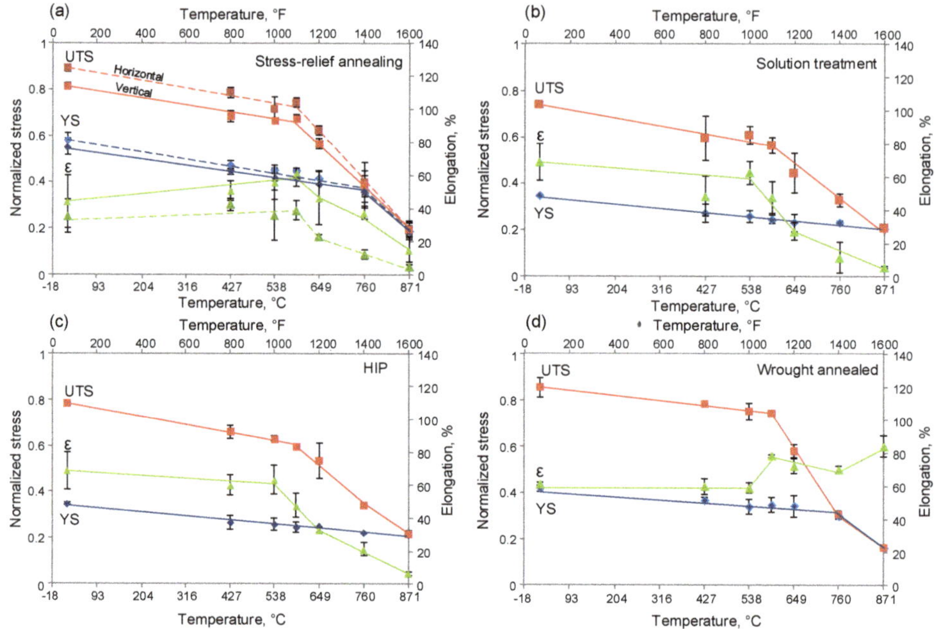

Figure 10. Mechanical properties of the (**a**) SR, (**b**) ST, (**c**) HIP, and (**d**) wrought annealed IN625 alloys as a function of temperature.

Regarding the mechanical properties overall, all specimens exhibited a decrease in the YS and UTS values as the test temperature increased, due to enhanced plastic flow causing tensile deformation at relatively lower stresses (Figure 5). However, their elongations to failure evolved differently, as discussed further.

Tensile strength characteristics (YS, UTS): The results show that at room temperature, the SR specimens exhibited the highest mechanical strengths (YS and UTS) (Figure 10a) and their mechanical strength characteristics exceeded those reported in the ASTM F3056-14 standard (min. 275 MPa (YS) and min. 485 MPa (UTS)) and EOS datasheet (min. 414 MPa (YS) and min. 827 MPa (UTS)) for LPBF IN625 alloy. Regarding other studies, the mechanical properties at room temperature of the SR specimens obtained in this study were comparable to those reported in References [13,19,21,38] for the as-built LPBF IN625 alloy. However, the high mechanical strength of the SR specimens came with a significant orientation dependency in the mechanical characteristics due to a strong microstructure anisotropy, which was inherited from the as-built material [16]. Thus, a finer microstructure in the build plane resulted in higher YS and UTS values of the horizontal specimens as compared to their vertical counterparts. By contrast, the ST and HIP specimens containing equiaxed grains and a random texture manifested build-orientation-independent behavior, but lower mechanical strength characteristics. The equiaxed grain structure and reduced mechanical strength characteristics after post-treatments performed at temperatures higher than 1100 °C have also been reported in References [13,21].

At elevated temperatures of up to 593 °C, the YS and UTS values decreased continuously with increasing temperature for all the tested specimens. The orientation dependency and the high mechanical strength characteristics of the SR specimens were preserved at these temperatures. The YS and UTS values of the ST and HIP specimens were still slightly lower than those of the wrought annealed alloy.

With a further temperature increase to up to 760 °C, a rapid decrease in mechanical strength characteristics was observed for the SR specimens, which was caused by dynamic recrystallization [24].

As a result, no difference was observed between the YS and UTS values. The ST- and HIP-treated specimens were capable of maintaining their structural strength to up to 760 °C, and their UTS values became comparable to those of the SR specimens. For the ST- and HIP-treated specimens, dynamic recrystallization took place at 871 °C. At this temperature, the YS and UTS values of all the tested specimens were comparable.

Elongation to failure: From RT to intermediate temperatures (538 °C), the elongations to failure increased for all the specimens. However, the peak in elongations was observed at 593 °C for the SR specimens, and at 538 °C for the ST and HIP specimens. Note that for the wrought annealed alloy, the elongation always increased with an increase in testing temperature (Figure 10d). At higher temperatures, the LPBF alloy manifested a significant reduction in elongation. Such a significant decrease in elongation of the LPBF alloy at 760 °C was accompanied by changes in fracture pattern due to weakening of the grain boundaries at elevated temperatures (ductile/brittle transition) [24]. Moreover, the intergranular cracking mode led to a significant increase in the orientation dependency for the SR specimens in the 593–871 °C temperature range. The elongated grains oriented along the axis of testing were responsible for higher elongations of the vertical SR specimens as compared to their horizontal counterparts. After the ST and HIP treatments, the grains became equiaxed and their mean size increased, resulting in elongations to failure higher than for the horizontal SR, but lower than for the vertical SR specimens.

The elevated temperature embrittlement observed in this study is a known phenomenon for many Ni-based alloys. The reasons for this behavior are subject to discussion [22]. In particular, the embrittlement has been attributed to intergranular precipitates, grain boundary shearing or sliding, gas phase embrittlement, dynamic strain aging, grain boundary segregation, and glide plane decohesion [22]. For IN625 alloy, this phenomenon is mostly associated with precipitates (carbides, $M_{23}C_6$ and M_6C, and δ phase) distributed along grain boundaries [2,24,39,40]. More specifically, according to Reference [24], cracks preferentially propagate along grain boundaries containing M_6C carbides.

Finally, it should be mentioned that the mechanical behaviors of the ST- and HIP-treated specimens were similar (Figure 10b,c). While the YS and UTS values of the ST specimens were slightly higher, their elongations were slightly lower than those of the HIP specimens.

High-temperature creep properties. It was found that as well as the tensile properties, the SR-annealed alloy exhibited orientation dependency of its creep behavior: vertically built specimens showed higher lifetimes and lower rupture strains compared to their horizontally built counterparts (Figures 7 and 8). Despite the high lifetimes of the vertical SR specimens as compared to the vertical ST and wrought specimens (Figure 11), the anisotropy in mechanical properties of the SR LPBF IN625 alloy makes its behavior unpredictable. Thus, heat treatment including only stress-relief annealing is not recommended for the practical use of LPBF IN625 alloy.

Regarding the ST-treated LPBF alloy, at the same creep-to-yield stress ratio, the ST specimens exhibited significantly improved creep properties as compared to the reference wrought annealed alloy. However, in the "absolute creep stress-rupture time" diagram (Figure 11), at first sight, the ST specimens still had an advantage over the wrought annealed alloy, especially at intermediate and low stresses (≤200 MPa). Using the "strain–time" creep curves, the steady creep rates were measured and are collected in Figure 12. It was clearly observed that the higher the applied stress, the higher the steady creep rate, which led to shorter lifetimes for the SR, ST, and wrought alloys, and a higher fraction of the intergranular fracture areas for the SR and ST alloys. However, it was seen that under the same creep stress of ≥150 MPa, the steady creep rates for the wrought and SR-treated LPBF alloys were higher than that for the ST-treated LPBF alloy.

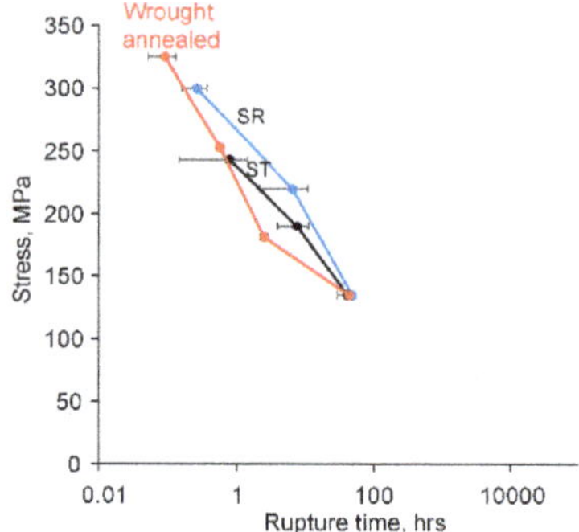

Figure 11. Applied stress versus creep rupture time at 760 °C for the SR, ST, and wrought annealed IN625 alloys.

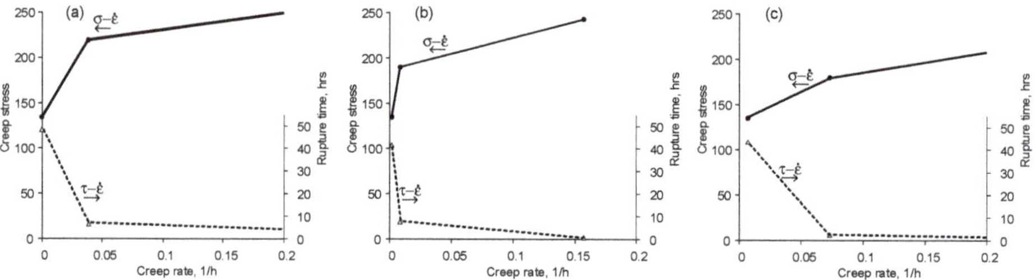

Figure 12. Creep stresses (σ) and mean rupture times (τ) vs. mean steady creep rates ($\dot{\varepsilon}$) for the (**a**) SR, (**b**) ST, and (**c**) wrought annealed IN625 alloys.

In general, creep behavior depends strongly on the metallurgical structure, i.e., the grain and particle sizes, the concentration of the alloying elements, and the creep conditions: stress and temperature. All the structure characteristics mentioned will affect the structural stability against vacancies and dislocations mobility, grain boundary diffusion, and, consequently, crack initiation and propagation at high temperatures under creep stresses [41]. In this study, for example, concentrations of Fe and C in the LPBF alloy were lower than in the wrought alloy. Note that according to the data of Heubner [42], elevated contents of Fe and C in IN625 negatively affect the creep behavior, especially at high temperatures.

Ni-based superalloys are polycrystalline and multiphase materials, and as such, correlating their properties with structural features is not straightforward. In particular, regarding the grain size influence, some data for IN718 demonstrate that an increase in grain size reduced the total creep rupture time, regardless of whether their grain boundaries are clean or intensively decorated with δ precipitates (67%) [43]. However, in classic cases, the trend is just the opposite: the coarser the structure, the greater the creep resistance [41]. A possible explanation for these discrepancies is probably related to the testing methodology: the same stress would create more severe creep conditions for coarse-grained structures with lower strength characteristics (YS, UTS) than for fine-grained structures with higher strength characteristics.

It has also been shown that the greater the density of δ precipitates at IN718 alloy grain boundaries, the shorter the total creep time, but when this density exceeds 45%, the creep time increases rapidly [43]. It has been suggested that with an increase in the density of precipitates at grain boundaries, the formation of wedge cracks at triple points can be delayed to the stage where creep voids around precipitates situated far from the triple edges are able to grow into unstable cracks. Thus, the beginning of the tertiary stage, characterized by rapid crack propagation, is delayed. Note, however, that in this study, all the creep tests were carried out at the same stresses, and thus, the variations in the mechanical strength of alloys with different precipitation densities were not taken into account.

A correlation between the creep rate (Figure 7) and the fractography (Figure 9) observations was be noted. For example, the specimens of the wrought annealed alloy accumulated a significant elongation to failure during their final (tertiary) creep stage, which was reflected by a dimpled fracture surface. At the same time, the ST specimens manifested a long, steady creep stage related to grain boundary sliding, void formation, and crack initiation, which was reflected by fast intergranular/transgranular crack propagation to failure.

Since the aim of the present article was mainly to present the tensile and creep properties of LPBF IN625 alloy, the authors will concentrate more efforts on establishing a correlation between the structural features and the functional properties in the next publication. However, to summarize, the SR-treated IN625 alloy contained needle-like δ phase and globular precipitates of M_6C carbides on grain boundaries [24], whereas the grain structures and phase states of the ST- and HIP-treated alloys were comparable. It has been previously shown that HIP dissolves the δ phase precipitates, forms MC, and homogenizes initially anisotropic SR-structure [24]. Thus, the presence of a significant quantity (that should be evaluated) of intergranular δ phase precipitates in vertical SR specimens reinforces grain boundaries against the sliding and formation of voids under stresses. After ST, in the case of an equiaxed grain structure with inter- and transgranular carbides, the creep lifetime is less significant, but the ST structure still has advantages over the wrought annealed specimen.

Summary. It was seen that the SR-treated LPBF alloy exhibited the highest anisotropy, i.e., build orientation dependency, and the least predictable mechanical behavior, as compared to its ST- and HIP-treated counterparts. Therefore, the SR alloy is the least safe material for practical use, especially at elevated temperatures. The ST and HIP treatments improved the alloy homogeneity and provided isotropic properties, thus making these alloys more application-safe. The mechanical strength characteristics of the LPBF alloy after the ST and HIP treatments satisfied the ASTM B444 standard requiring high strength in the 25–593 °C (68–1100 °F) temperature range.

Note that while the HIP alloy can be seen as a material with more uniform mechanical characteristics and improved ductility over the ST-treated alloy, this advantage came at the expense of the lower strength characteristics.

Furthermore, at T ≥ 650 °C, special attention must be paid to the time-dependent properties (creep). It was seen that although the LPBF alloy manifested much lower static ductility at these temperatures, it offered significantly longer rupture times under stresses of <200 MPa, as compared to its wrought annealed counterpart (Figure 13).

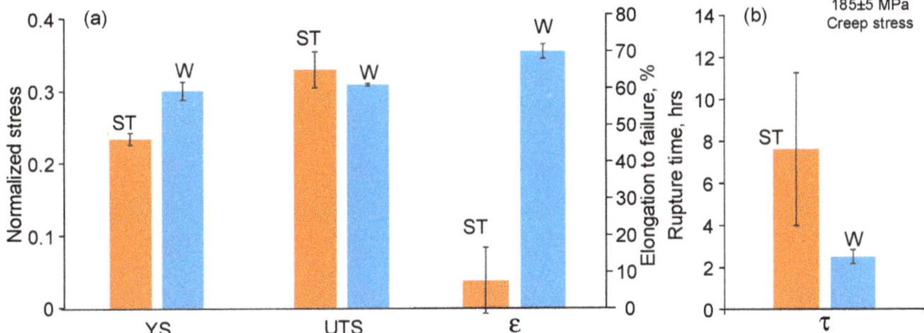

Figure 13. Comparison of the high temperature mechanical properties (760 °C, 1400° F) of the ST LPBF and wrought annealed IN625 alloys: (**a**) strength (YS, UTS) and elongation to failure (ε), and (**b**) creep rupture time (τ) under a creep stress of 185 ± 5 MPa.

5. Conclusions

1. After SR annealing, the microstructure of LPBF IN625 alloy is anisotropic, characterized by elongated grains mainly oriented along the build direction. ST and HIP treatments lead to the formation of a uniform microstructure.

2. SR annealing provides anisotropic mechanical properties and low elongations, and the highest tensile strength in the 25–593 °C temperature range. The ST and HIP treatments improved ductility, but this improvement was obtained at the expense of lower mechanical strength characteristics.

3. At elevated temperatures (649–871 °C), the LPBF specimens exhibited low ductility as compared to the wrought annealed alloy. This decrease in ductility was accompanied by a transition from transgranular to intergranular cracking mode.

4. The ST-annealed LPBF alloy exhibited significantly improved creep properties at 760 °C as compared to the wrought annealed alloy, especially at intermediate and low stresses (≤200 MPa).

Author Contributions: The work plan was developed by A.K. and K.I to meet the study objectives defined by V.B and S.T.; Specimens' design, testing and data analysis were performed by A.K. and K.I.; Original draft was prepared by A.K. and K.I; V.B. and S.T. contributed to results interpretation, manuscript review and editing.

Funding: This research was founded by NSERC (Natural Sciences and Engineering Research Council of Canada), CRIAQ (Consortium de Recherche et d'Innovation en Aérospatiale du Québec), Pratt & Whitney Canada, Fusia, Thermetco and FZ Engineering (Project MANU 1625).

Conflicts of Interest: The authors declare no conflict of interest.

References

1. Eiselstein, H.L.; Tillack, D.J. *The Invention and Definition of Alloy 625, Superalloys 718, 625 and Various Derivatives*; Loria, E.A., Ed.; TMS: Pittsburgh, PA, USA, 1991; pp. 1–14.
2. Shankar, V.; Valsan, M.; Rao, K.B.S.; Mannan, S.L. Effects of temperature and strain rate on tensile properties and activation energy for dynamic strain aging in alloy 625. *Met. Mater. Trans. A* **2004**, *35*, 3129–3139. [CrossRef]
3. Liu, D.; Zhang, X.; Qin, X.; Ding, Y. High-temperature mechanical properties of Inconel-625: Role of carbides and delta phase. *Mater. Sci. Technol.* **2017**, *33*, 1610–1617. [CrossRef]
4. *MIL-HDBK-5H: Metallic Materials and Elements for Aerospace Vehicle Structures*; Military Handbook; U.S. Department of Defense: Washington, DC, USA, 1998.
5. Deel, O. *Mechanical Property Data Inconel 625*; Battelle Memorial Institute, Defense Technical Information Center: Columbus, OH, USA, 1971.
6. Kohler, M. *Effect of the Elevated-Temperature-Precipitation in Alloy 625 on Properties and Microstructure, Superalloys 718, 625 and Various Derivatives*; Loria, E.A., Ed.; The Minerals, Metals and Materials Society: Pittsburgh, PA, USA, 1991; pp. 363–374.
7. Debroy, T.; Wei, H.; Zuback, J.; Mukherjee, T.; Elmer, J.; Milewski, J.; Beese, A.; Wilson-Heid, A.; De, A.; Zhang, W. Additive manufacturing of metallic components—Process, structure and properties. *Prog. Mater. Sci.* **2018**, *92*, 112–224. [CrossRef]
8. Herzog, D.; Seyda, V.; Wycisk, E.; Emmelmann, C. Additive manufacturing of metals. *Acta Mater.* **2016**, *117*, 371–392. [CrossRef]
9. Bhavar, V.; Kattire, P.; Patil, V.; Khot, S.; Gujar, K.; Singh, R. A review on powder bed fusion technology of metal additive manufacturing. In Proceedings of the 4th International Conference and Exhibition on Additive Manufacturing Technologies-AM-2014, Banglore, India, 1–2 September 2014; pp. 1–2.
10. Donachie, M.J.; Donachie, S.J. *Superalloys a Technical Guide*, 2nd ed.; ASM International: Materials Park, OH, USA, 2002.
11. Amato, K.; Gaytan, S.; Murr, L.; Martinez, E.; Shindo, P.; Hernandez, J.; Collins, S.; Medina, F.; Murr, L. Microstructures and mechanical behavior of Inconel 718 fabricated by selective laser melting. *Acta Mater.* **2012**, *60*, 2229–2239. [CrossRef]
12. Li, C.; White, R.; Fang, X.; Weaver, M.; Guo, Y. Microstructure evolution characteristics of Inconel 625 alloy from selective laser melting to heat treatment. *Mater. Sci. Eng. A* **2017**, *705*, 20–31. [CrossRef]

13. Amato, K.N.; Hernandez, J.; Murr, L.E.; Martinez, E.; Gaytan, S.M.; Shindo, P.W.; Collins, S. Comparison of Microstructures and Properties for a Ni-Base Superalloy (Alloy 625) Fabricated by Electron and Laser Beam Melting. *J. Mater. Sci. Res.* **2012**, *1*, 3–41.

14. Li, Y.; Gu, D. Parametric analysis of thermal behavior during selective laser melting additive manufacturing of aluminum alloy powder. *Mater. Des.* **2014**, *63*, 856–867. [CrossRef]

15. Shiomi, M.; Osakada, K.; Nakamura, K.; Yamashita, T.; Abe, F. Residual Stress within Metallic Model Made by Selective Laser Melting Process. *CIRP Ann.* **2004**, *53*, 195–198. [CrossRef]

16. Kreitcberg, A.; Brailovski, V.; Turenne, S. Effect of heat treatment and hot isostatic pressing on the microstructure and mechanical properties of Inconel 625 alloy processed by laser powder bed fusion. *Mater. Sci. Eng. A* **2017**, *689*, 1–10. [CrossRef]

17. Mostafa, A.; Rubio, I.P.; Brailovski, V.; Jahazi, M.; Medraj, M. Structure, Texture and Phases in 3D Printed IN718 Alloy Subjected to Homogenization and HIP Treatments. *Metals* **2017**, *7*, 196. [CrossRef]

18. Ferrer, L.; Pieraggi, B.; Uginet, J.F. *Microstructure Evolution during Thermomechanical Processing of Alloy 625, Superalloys 718, 625 and Various Derivatives*; Loria, E.A., Ed.; The Minerals, Metals and Materials Society: Pittsburgh, PA, USA, 1991; pp. 217–228.

19. Brown, C.U.; Jacob, G.; Stoudt, M.; Moylan, S.; Slotwinski, J.; Donmez, A. Interlaboratory study for nickel alloy 625 made by laser powder bed fusion to quantify mechanical property variability. *J. Mater. Eng. Perform.* **2016**, *25*, 3390–3397. [CrossRef] [PubMed]

20. Hack, H.; Link, R.; Knudsen, E.; Baker, B.; Olig, S. Mechanical properties of additive manufactured nickel alloy 625. *Addit. Manuf.* **2017**, *14*, 105–115. [CrossRef]

21. Marchese, G.; Lorusso, M.; Parizia, S.; Bassini, E.; Lee, J.-W.; Calignano, F.; Manfredi, D.; Terner, M.; Hong, H.-U.; Ugues, D.; et al. Influence of heat treatments on microstructure evolution and mechanical properties of Inconel 625 processed by laser powder bed fusion. *Mater. Sci. Eng. A* **2018**, *729*, 64–75. [CrossRef]

22. Zheng, L.; Schmitz, G.; Meng, Y.; Chellali, R.; Schlesiger, R. Mechanism of Intermediate Temperature Embrittlement of Ni and Ni-based Superalloys. *Crit. Rev. Solid State Mater. Sci.* **2012**, *37*, 181–214. [CrossRef]

23. Murr, L.E.; Martinez, E.; Gaytan, S.M.; Ramirez, D.A.; Machado, B.I.; Shindo, P.W.; Martinez, J.L.; Medina, F.; Wooten, J.; Ciscel, D.; et al. Microstructural Architecture, Microstructures, and Mechanical Properties for a Nickel-Base Superalloy Fabricated by Electron Beam Melting. *Met. Mater. Trans. A* **2011**, *42*, 3491–3508. [CrossRef]

24. Kreitcberg, A.; Brailovski, V.; Turenne, S. Elevated temperature mechanical behavior of IN625 alloy processed by laser powder-bed fusion. *Mater. Sci. Eng. A* **2017**, *700*, 540–553. [CrossRef]

25. Davis, J.R. *ASM Specialty Handbook: Nickel, Cobalt, and Their Alloys*; ASM International: Geauga County, OH, USA, 2000; p. 442.

26. Kuo, Y.-L.; Horikawa, S.; Kakehi, K. Effects of build direction and heat treatment on creep properties of Ni-base superalloy built up by additive manufacturing. *Scr. Mater.* **2017**, *129*, 74–78. [CrossRef]

27. Kuo, Y.-L.; Nagahari, T.; Kakehi, K. The Effect of Post-Processes on the Microstructure and Creep Properties of Alloy718 Built Up by Selective Laser Melting. *Materials* **2018**, *11*, 996. [CrossRef]

28. Xu, Z.; Murray, J.W.; Hyde, C.J.; Clare, A.T. Effect of post processing on the creep performance of laser powder bed fused Inconel 718. *Addit. Manuf.* **2018**, *24*, 486–497. [CrossRef]

29. Xu, Z.; Hyde, C.J.; Tuck, C.; Clare, A.T. Creep behavior of Inconel 718 processed by laser powder bed fusion. *J. Mater. Process. Technol.* **2018**, *256*, 13–24. [CrossRef]

30. Pröbstle, M.; Neumeier, S.; Hopfenmüller, J.; Freund, L.P.; Niendorf, T.; Schwarze, D.; Göken, M. Superior creep strength of a nickel-based superalloy produced by selective laser melting. *Mater. Sci. Eng. A* **2016**, *674*, 299–307. [CrossRef]

31. Chang, S.-H.; Lee, S.-C.; Tang, T.-P.; Ho, H.-H. Influences of Soaking Time in Hot Isostatic Pressing on Strength of Inconel 718 Superalloy. *Mater. Trans.* **2006**, *47*, 426–432. [CrossRef]

32. "Special Metals", Inconel Alloy 625. Available online: www.specialmetals.com/assets/smc/documents/alloys/inconel/inconel-alloy-625.pdf (accessed on 31 May 2019).

33. "Special Metals", Inconel Alloy 625. Available online: https://www.haraldpihl.com/globalassets/pdf/033_inconel-alloy-625lcf.pdf (accessed on 31 May 2019).

34. Das, S.; Wohlert, M.; Beaman, J.J.; Bourell, D.L. Direct Selective Laser Sintering of high performance metals for containerless HIP. *Adv. Powder Metall. Part. Mater.* **1997**, *3*, 81–90.

35. Carlson, R.G. *Cast 625 Hot Isostatic Pressing (HIP) Parameters—A Statistically Designed Study, Superalloys 718, 625 and Various Derivatives*; Loria, E.A., Ed.; The Minerals, Metals and Materials Society: Pittsburgh, PA, USA, 1991; pp. 97–106.

36. ASTM. *Standard Specification for Additive Manufacturing Nickel Alloy (UNS N06625) with Powder Bed Fusion*; ASTM International: West Conshohocken, PA, USA, 2014.

37. Kohler, M.; Heubner, U. *The Effect of Final Heat Treatment and Chemical Composition on Sensitization, Strength and Thermal Stability of Alloy 625. Superalloys 718,625,706 and Various Derivatives*; Loria, E.A., Ed.; The Minerals, Metals & Materials Society: Pittsburgh, PA, USA, 1997; pp. 795–803.

38. Yadroitsev, I.; Thivillon, L.; Bertrand, P.; Smurov, I.; Bertrand, P. Strategy of manufacturing components with designed internal structure by selective laser melting of metallic powder. *Appl. Surf. Sci.* **2007**, *254*, 980–983. [CrossRef]

39. Vernot-Loier, C.; Cortial, F. *Influence of Heat Treatments on Microstructure, Mechanical Properties and Corrosion Behavior of Alloy 625 Forged Rod, Superalloys 718, 625 and Various Derivatives*; Loria, E.A., Ed.; The Minerals, Metals and Materials Society: Pittsburgh, PA, USA, 1991; p. 409.

40. Cortial, F.; Corrieu, J.M.; Vernot-Loier, C. *Heat Treatments of Weld Alloy 625: Influence on the Microstructure, Mechanical Properties and Corrosion Resistance, Superalloys 718, 625, 706 and Various Derivatives*; Loria, E.A., Ed.; The Minerals, Metals and Materials Society: Pittsburgh, PA, USA, 1994; p. 859.

41. Reed, R.C. *The Superalloys: Fundamentals and Applications*; Cambridge University Press: New York, NY, USA, 2006.

42. Heubner, U.; Kohler, M. *Effect of Carbon Content and Other Variables on Yield Strength, Ductility and Creep Properties of Alloy 625, Superalloys 718,625,706 and Various Derivatives*; Loria, E.A., Ed.; The Minerals, Metals & Materials Society: Pittsburgh, PA, USA, 1994; pp. 479–488.

43. Chen, W.; Chaturvedi, M. Dependence of creep fracture of inconel 718 on grain boundary precipitates. *Acta Mater.* **1997**, *45*, 2735–2746. [CrossRef]

Journal of
*Manufacturing and
Materials Processing*

Article

Finite Element Modeling of Orthogonal Machining of Brittle Materials Using an Embedded Cohesive Element Mesh

Behrouz Takabi and Bruce L. Tai *

Department of Mechanical Engineering, Texas A&M University, 3123 TAMU, College Station, TX 77843, USA; btakabi@tamu.edu
* Correspondence: btai@tamu.edu; Tel.: +1-979-458-9888

Received: 6 April 2019; Accepted: 29 April 2019; Published: 2 May 2019

Abstract: Machining of brittle materials is common in the manufacturing industry, but few modeling techniques are available to predict materials' behavior in response to the cutting tool. The paper presents a fracture-based finite element model, named embedded cohesive zone–finite element method (ECZ–FEM). In ECZ–FEM, a network of cohesive zone (CZ) elements are embedded in the material body with regular elements to capture multiple randomized cracks during a cutting process. The CZ element is defined by the fracture energy and a scaling factor to control material ductility and chip behavior. The model is validated by an experimental study in terms of chip formation and cutting force with two different brittle materials and depths of cut. The results show that ECZ–FEM can capture various chip forms, such as dusty debris, irregular chips, and unstable crack propagation seen in the experimental cases. For the cutting force, the model can predict the relative difference among the experimental cases, but the force value is higher by 30–50%. The ECZ–FEM has demonstrated the feasibility of brittle cutting simulation with some limitations applied.

Keywords: orthogonal cutting; brittle materials; cohesive elements

1. Introduction

Machining of brittle materials such as ceramics, rocks, composites, and bones is common in aerospace/automotive industries and the medical field [1]. Although efforts [2–6] have been made to model machining of fiber-reinforced composite materials for predicting brittle failure, there is not a generalized method that can successfully and efficiently emulate the physics behind brittle cutting—the rapid and randomized crack initiation and propagation upon tool-workpiece contact. Unlike ductile material cutting, which is dominated by shear deformation across the shear plane, brittle material cutting is driven by fractures. Finite element method (FEM) has been widely used to simulate ductile material machining (e.g., metals) using the Johnson–Cook plasticity model for cutting forces and chip formation [7–9]. However, FEM has not yet been successfully applied to brittle materials because of the difficulty of capturing numerous and unpredictable cracks at the same time. Technically, FEM needs an extremely fine mesh to simulate stress concentration and consequent element failure at each time increment, which is not practical due to a high computational cost.

Researchers have tried to apply mesh-free methods such as smooth particle hydrodynamics (SPH) to cutting simulation because they do not require a gridded domain and can handle large deformation [10]. However, there are discrepancies among the published works, especially on damage definition. Takabi et al. [11] investigated SPH in orthogonal cutting and showed the uncertainty of damage due to particles losing connection to each other (i.e., the natural separation), which can drastically change the outcome. Also, particle separation is not determined by the fracture toughness but the material strength. Therefore, mesh-free methods are not considered an ideal approach for brittle materials cutting.

To deal with fracture problems, the cohesive element has been developed for FEM, which forms the cohesive zone (CZ) in the model. The cohesive zone concept links the microstructural failure mechanism to the continuum fields [12]. A CZ element can begin to separate based on the strain energy release rate, which is often defined by a traction–displacement relationship. The cohesive zone–finite element method (CZ–FEM) has been a useful tool for investigation of interfacial fracture problems, such as crack tip propagation, the adhesive strength between two materials, and modeling of composite delamination. CZ–FEM has been used to solve machining problems of composites and ceramics, though not many. Rao et al. [2] simulated the orthogonal cutting of unidirectional carbon fiber-reinforced polymer and glass fiber-reinforced polymer composites using CZ between the fibers and matrix. They used a 2D plane strain model and zero-thickness cohesive elements to enable fiber detachment when the interfacial energy exceeds the threshold defined by an exponential traction–displacement relationship. Umer et al. [3] used CZ–FEM to simulate metal matrix composite machining. They modeled the orthogonal machining of SiC particle-reinforced aluminum-based metal matrix composites by placing CZ elements between the particles and the matrix. A bilinear traction–displacement profile was used for CZ elements with zero thickness. Dong and Shin [13] developed a multi-scale model for simulating the machining of alumina ceramics in laser-assisted machining. Zero-thickness CZ was assigned around the ceramic grain boundaries, and the traction–displacement profile was determined based on a separate molecular dynamics (MD) simulation. Note that CZ is often modeled as zero thickness because it is an imaginary interface inside the material in these cases, unlike physical adhesives.

In the above-mentioned CZ–FEM works, the CZ elements are placed either at known interfaces or paths as a pre-determined condition where cracks will initiate and propagate [14]. Therefore, CZ–FEM does not seem possible for a homogenous, flaw-free brittle material in which potential cracking path cannot be defined. To address this issue, the current study proposes using a CZ mesh together with a regular element mesh to enable a network of potential cracks. A zero-thickness CZ element is embedded between regular elements. In other words, this CZ mesh will force the material to fail between elements instead of within an element. This modified CZ–FEM is named embedded cohesive zone–finite element method (ECZ–FEM). The ECZ–FEM for brittle machining is developed and validated in this paper using the commercial FEM software ABAQUS.

2. Finite Element Model Setup

This section presents the overall configuration of ECZ–FEM, step-by-step procedures to construct the model, and the required modification for material properties. The model introduced here is built based on the corresponding orthogonal cutting experiment.

2.1. Model Configuration

A two-dimensional orthogonal, plane strain cutting model is configured in ABAQUS (version 6.14-2), as illustrated in Figure 1. There are two main sections in this model. The top section (named the chip layer) is the ECZ domain where CZ elements are embedded all around the main elements. The bottom section is the regular finite element domain without CZ elements. This configuration saves computational time compared to a fully embedded CZ model since the bottom layer is not directly involved in the tool–workpiece interaction. Instead, the bottom layer provides compliance to the material during cutting. To ensure the stress continuity, the nodes on both sides of the interface must be merged or tied with all degrees of freedom. For this reason, the mesh sizes on both sections must match to have a perfect node-to-node alignment.

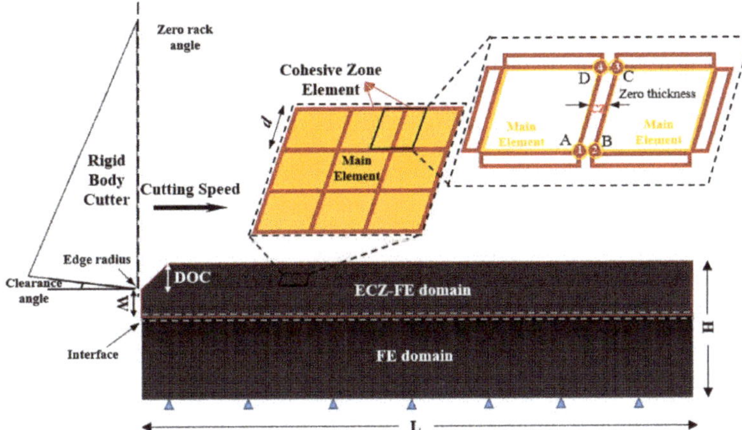

Figure 1. Schematic of embedded cohesive zone–finite element method (ECZ–FEM) model configuration, boundary conditions, and element arrangements.

Table 1 shows the actual model dimensions used for two depths of cut (DOC), 0.1 mm and 0.3 mm. The boundary of the bottom layer is fixed in both translational directions (X and Y). The element size, d, is set at 0.01 mm. The bottom layer is meshed structurally with brick elements, while for the top layer, the elements are tilted by 45°. The inclined elements are necessary because the maximum shear stress to fracture the material is expected to be around 45° based on the Merchant's Circle [15]. This configuration can avoid numerical instability when no CZ mesh aligns with the preferred fracture direction. The main elements are the four-node plane strain elements CPE4R, and the CZ elements are the four-node two-dimensional cohesive elements COH2D4. To embed zero-thickness CZ elements, all elements and nodes of the chip layer need to be assigned through the input file directly because each CZ element shares nodes with adjacent two main elements, as shown in Figure 1. The CZ element is defined by nodes ABCD, in which A and D belong to the element on the left side (identical to Nodes 1 and 4), while B and C belong to the right side (identical to Nodes 2 and 3). Since these two pairs of nodes are overlaid geometrically, they cannot be identified from the graphic user interface. The meshing process is automatized by a separate MATLAB code.

Table 1. The model dimensions and depths of cut (DOC) used.

	DOC (mm)	L (mm)	H (mm)	W (mm)
Case 1	0.1	2	0.5	0.1
Case 2	0.3	5	0.85	0.1

A complete mesh is imported to ABAQUS/EXPLICIT to set up other boundary conditions. The plane strain thickness of 3 mm is also applied to the model to be consistent with the thickness of the actual sample. The cutting tool is modeled as a rigid body with a constant speed at 10 m/min to match with the experiment. The tool has a rake angle of zero, a clearance angle of 7°, and an edge radius of 11 μm.

2.2. Damage Criteria

To apply the ECZ–FEM to a brittle cutting process, the material properties of the main and CZ elements and their damage criteria are defined separately despite being within the same entity. Assuming an isotropic, brittle material, the main element is defined by the modulus of elasticity (E), Poisson's ratio (μ), the ultimate strength (σ_u), and damage criteria of the material. Although the model is meant to impart fracture-based failure on the CZ mesh, the main element should still allow failing to avoid excessive element distortion when no fracture occurs. For this reason, the damage to the main elements is defined by an initiation at the ultimate strength followed by a progressive damage evolution by the Hillerborg's fracture energy theory. The total energy required to completely degrade the element after the damage initiation is G_f, which can be calculated from the material's fracture toughness K_c by Equation (1):

$$G_f = \left(\frac{1 - v^2}{E}\right) K_c^2. \tag{1}$$

The degradation is in a linear manner [16], such that

$$D = \frac{\bar{u}}{\bar{u}_f}, \tag{2}$$

where $\bar{u}$ is the equivalent element displacement after the damage initiation; $\bar{u}_f$ represents the equivalent displacement at failure. The displacement at failure is calculated by

$$\bar{u}_f = \frac{2G_f}{\sigma_u}, \tag{3}$$

where σ_u represents the ultimate stress. These are standard steps to simulate material failure for metal cutting [16]. It should be emphasized that this damage definition for the main element is to ensure the model stability by avoiding excessive element distortion.

The properties associated with the CZ elements embedded in the chip layer are defined differently. The cohesive zone is a mathematical approach in which the work is done to overcome the energy needed to open a crack. This work can be described by a traction–displacement relationship, t-δ, as seen in Figure 2. Damage initiation is related to the interfacial strength (i.e., the maximum traction t_c) on the traction–displacement relation, and the area under the relation represents the fracture energy, G_f, as defined in Equation (1).

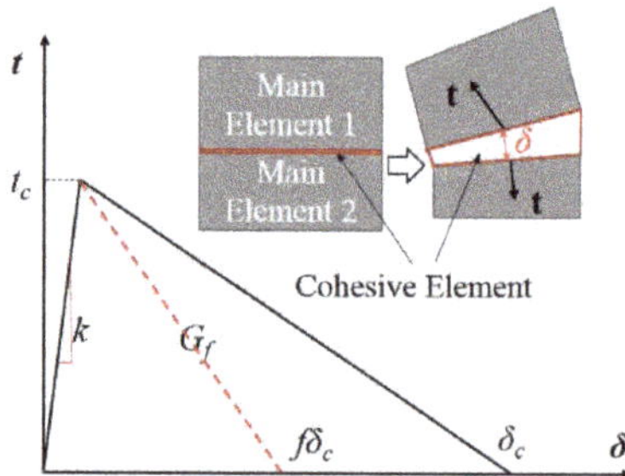

Figure 2. Bilinear traction–displacement (t-δ) model for the cohesive element.

In this study, a bilinear traction–separation law is adopted along with the mixed-mode progressive damage. The maximum traction t_c should be equal or less than the strength of the material to be able to fail, while a lower strength can improve the convergence rate of the solution. In general, the variations of the maximum strength do not have a strong influence on the results [12]. Hence, the 80% ultimate stress is selected here. The initial stiffness k should be large enough to ensure the continuum between the two adjacent bulk elements, but small enough to avoid numerical issues such as spurious oscillations of the tractions in an element. Studies suggest that the initial stiffness of CZ elements can be calculated from Equation (4), which balances accuracy and simulation stability [12,17,18].

$$k = \alpha \frac{E}{d}, \tag{4}$$

where E is the bulk elasticity, d is the maximum element size, and α is taken as 1.

The maximum deflection of a CZ element δ_c is determined by given G_f and t_c, as shown in Figure 2. Therefore, the deflection can become relatively large compared to the element size when a fine mesh is used. A large deflection is infeasible since it increases the material ductility when a CZ mesh is embedded in the material, as shown in Figure 3. When the material is subject to stresses to deform, the original element size (d) will increase to ($d' + \delta$), which adds additional elongation δ to the material. Because of this limitation, a scaling factor (denoted as f) is introduced here to limit the maximum deflection of CZ elements, as shown by $f\delta_c$ in Figure 2, and thus to control the chip behavior. Chip behavior is a critical indicator as the cutting force can be affected by the number of cracks during cutting (i.e., work done vs. total fracture energy).

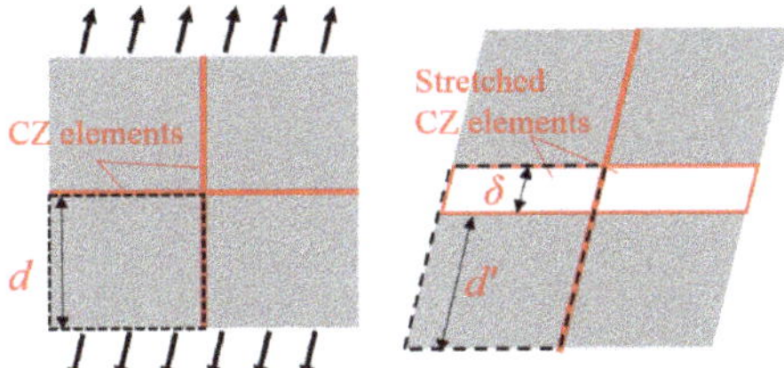

Figure 3. A schematic drawing to show unrealistic deformation due to the deflection of cohesive zone (CZ) elements.

When the deflection is scaled to control chip behavior, the fracture energy and, therefore, the cutting force will be scaled accordingly. Thus, the cutting force must be inversely scaled to represent the actual force. To properly apply this model with the scaling factor, the following assumptions are made. First, beyond the elastic deformation, no plastic deformation occurs in the material and all the force contributes to material removal. Second, the specific cutting energy (energy required to remove a unit volume of material) is based solely on the fracture energy. Given constant cutting velocity v_f and material removal rate (MRR), the cutting force (F_c) will be linearly proportional to the specific cutting energy (p), as described in Equation (5),

$$F_c v_f = MRR \cdot p. \tag{5}$$

This implies that the cutting force is scaled linearly with the CZ element's fracture energy. This concept will be validated in the experimental study.

2.3. Other Material Properties

The brittle materials used for the experiment are two types of solid bone-mimetic materials made of high-density polyurethane (PU) foam (Sawbones, Vashon, WA, USA). This material provides consistent and uniform material properties; it is isotropic and does not require a large force to cut. It is ideal for the modeling purpose and experimental validation without extraneous variables such as vibration, impact shock, and heat. These two foams are named based on their densities, 30 and 40 pcf (pound per cubic foot), which equates to 480 kg/m^3 and 640 kg/m^3, respectively. The 30 pcf has the ultimate strength of 12 MPa and the elasticity modulus of 592 MPa; the 40 pcf has the strength of 19 MPa, and the modulus of 1000 MPa, respectively, based on the manufacturer provided data [19]. The fracture toughness, K_c, of these foams is obtained from a separate three-point bending experiment following ASTM D5045–93. The averaged K_c for the 30 pcf is 0.46 MPa.m$^{1/2}$ and that of 40 pcf is 1.13 MPa.m$^{1/2}$. The 40 pcf is stiffer and also tougher than the 30 pcf. Based on these properties, the original CZ element properties are calculated in Table 2 below. As seen, the allowed cohesive element deformations are both larger than the element itself (0.01 mm).

Table 2. The CZ element properties for the testing materials 30 pound per cubic foot (pcf) and 40 pcf.

Samples	t_c (N/mm^2)	k (N/mm^3)	G_f (N/mm)	δ (mm)
30 pcf	9.6	59,200	0.31	0.064
40 pcf	15.2	100,000	1.12	0.147

2.4. Scaling Factor

The scaling factor is necessary to control the maximum deflection of CZ elements and thus the material ductility. In the case of 30 pcf, the original CZ deflection goes up to 0.064 mm. With the adjacent element size being 0.01 mm, this allowable deflection is equivalent to a 600% additional elongation (0.064/0.01), which is unrealistic. Figure 4 shows four different scenarios when using the original G_f and scaled G_f that limits the deflection to be 0.00512 mm (51.2% elongation), 0.00128 mm (12.8% elongation), and 0.00032 mm (3.2% elongation), respectively. As seen in Figure 4a with the original G_f, the workpiece and elements experience excessive deformation. Many stretched CZ elements remain alive though the chip has been distorted significantly. Figure 4b shows small but consistent chips generated from the shear plane, which is similar to cutting of brittle metals like high carbon steels. Figure 4c begins to generate fragmented, irregular debris accompanied by dusty pieces, which can be similar to ceramic materials. Figure 4d shows a more extreme case, where the workpiece shatters upon the tool contact. From these simple examples, it can be seen that a fairly small scaling factor is needed in order to force the material to behave as brittle. Note that in the simulation no self-contact is employed because the elements are supposed to support each other via CZ elements. Self-contact is possible but will exponentially increase the computational load due to the larger number of surfaces involved in the contact algorithm.

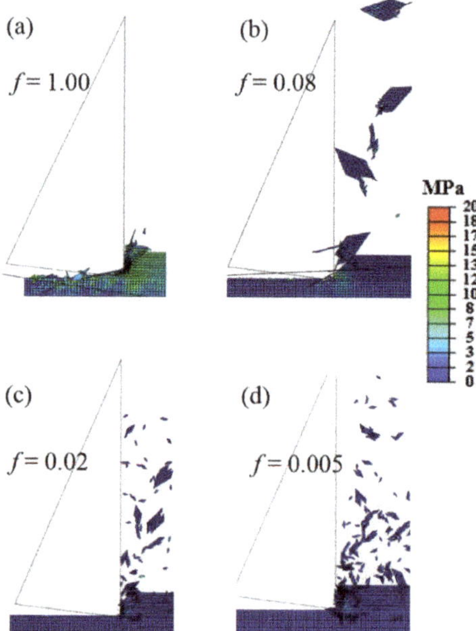

Figure 4. Material responses to the cutting tool with different scaling factors: (**a**) $f = 1$, (**b**) $f = 0.08$, (**c**) $f = 0.02$, and (**d**) $f = 0.005$. The stress is based on the testing material 30 pcf.

2.5. Sensitivity Study

Simulation output depends on the mesh configuration, such as the element size and the tilt angle. The element size of 0.01 mm was selected to compromise between the computational load and convergence. A smaller mesh size of 0.005 mm was compared to the 0.01 mm mesh using the 30 pcf case and showed a similar force magnitude and chip formation, but the computation could hardly proceed after a few steps due to a large number of elements and surfaces.

For the tilt angle, although 45° is the theoretically preferred cracking path, different angles were also tested at 0° (square mesh), 30°, and 60° to study the mesh sensitivity using the 30 pcf case with a scaling factor of 0.02. In the case of square mesh, the chip layer was sheared off without any cutting phenomenon due to the lack of fracture path around the theoretical shear angle. Results of the other cases are shown in Figure 5. Compared to the 45° mesh, the chip size is larger at 30° and smaller at 60°. Consequently, the cutting force is a little smaller (about 2%) at the 30° mesh and larger (about 20%) at the 60° mesh because the work done of cutting force is proportional to the number of fractured surfaces. Therefore, although the 45° mesh is recommended based on the shear angle, other mesh angles may also work but would produce different results. In any case, the scaling factor needs to be adjusted to match the chip behavior to the experiment for the best outcome.

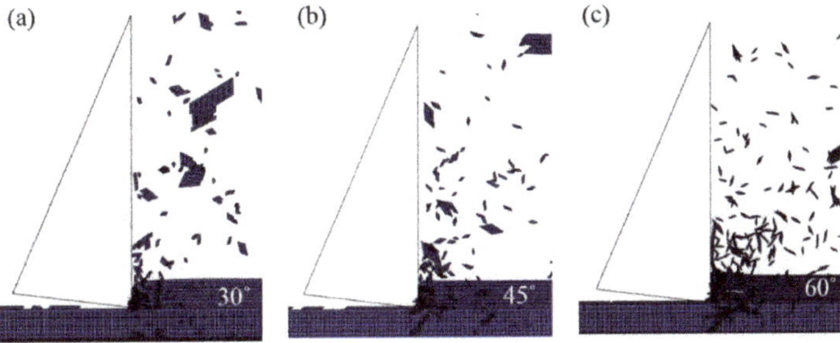

Figure 5. Material responses to the cutting tool with different tile angles: (**a**) 30 degrees, (**b**) 45 degrees, and (**c**) 60 degrees.

3. Experiment Setup for Model Validation

Orthogonal cutting is the basic cutting configuration for all machining processes. The essential geometrical parameters include rake angle, clearance angle and the depth of cut. In this experiment, the solid foams, the 30 and 40 pcf, are sectioned to a 20 × 30 × 3 mm testing sample. Each sample is hand-polished with the same grit size to ensure a smooth surface and uniform depth of cut. Figure 6 illustrates the experimental setup for the orthogonal cutting setup which consists of two linear actuators and a dynamometer for force measurement. The cutting tool is attached to the vertical linear actuator through a customized tool holder. The linear actuator (L70, Moog Animatics, Milpitas, CA, USA) is driven by a high-torque servo-motor to maintain a constant feed rate during cutting. The force dynamometer (Model 9272, Kistler, Winterthur, Switzerland) is used to capture high-speed or high-frequency force data up to 5 kHz. Data collection is performed via an amplifier, a shielded connector block, and a data acquisition device (PCIe-6321, National Instruments, Austin, TX, USA), along with a data recorder, LabVIEW, at 2 kHz sampling rate. The workpiece is fixed by a clamping system on the top of the dynamometer which is placed on the other linear slider to control the depth of cut for each test.

The cutting tool has a tungsten carbide substrate and a polycrystalline diamond (PCD) insert as a cutting edge, provided by Sandvik (Model TCMW16T304FLP-CD10). This PCD insert is extremely hard and minimizes any possible deformation or wear at the cutting edge. This cutting tool has a zero-rake angle and a clearance angle of 7°. The cutting edge radius is 11 µm, measured by a high-definition surface profiler (Alicona InfiniteFocus G4, Graz, Austria).

In this experiment, two depths of cut, 0.1 mm and 0.3 mm, are used to present common chip loads for a machining process. The cutting tool is moved at a constant velocity of 10 m/min to represent a machining condition. These parameters are applied to two specimens and repeated for four times each.

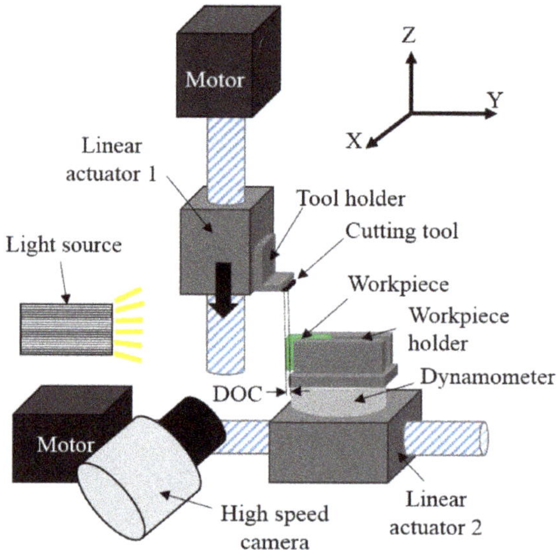

Figure 6. Schematic of the orthogonal cutting setup for model validation.

4. Simulation and Experiment Results

The simulation results are compared to the experiments in different depths of cut and material properties (30 pcf and 40 pcf) in this section.

4.1. Chip Formation

To find an appropriate scaling factor, a qualitative comparison of chip formation behavior against the experiment is conducted. In brittle materials, the chip can be generated in various forms, including dusty debris, fragmented and irregular pieces, or equal-sized small chips. Different scaling factors are tested until a similar chip behavior to the experiment is achieved or no obvious behavior difference can be observed. For this purpose, the initial guess for the scaling factor is recommended to be half of the element size (i.e., $\delta_c/d = 0.5$) to ensure the material brittleness. Then, a binary search method is used. If the current f does not show a good match, half of the value ($f/2$) will be investigated until the best fit is found or further improvement is not distinguishable.

Following the aforementioned procedure, the model calibration is performed for 30 pcf and DOC = 0.1 mm. Figure 7a shows the corresponding simulation results with a selected $f = 0.02$, which has a similar chip formation to that of the experiment. The simulation can capture the irregular chips of different sizes generated from the cutting zone. Then this scaling factor is also used to simulate the case of 0.3 mm DOC. The result is shown in Figure 7b. A larger DOC tends to generate bigger chips surrounded by small debris as compared to the case of 0.1 mm DOC. Consistently, the experiment also sees much bigger or clustered pieces when DOC increases to 0.3 mm. The results of 30 pcf with the selected scaling factor show qualitative agreement between the model and experiment in terms of chip behavior. Chip sizes of simulation and experiment do not match exactly due to the limited observation window and material uncertainty, but the difference is in the order of sub-mm.

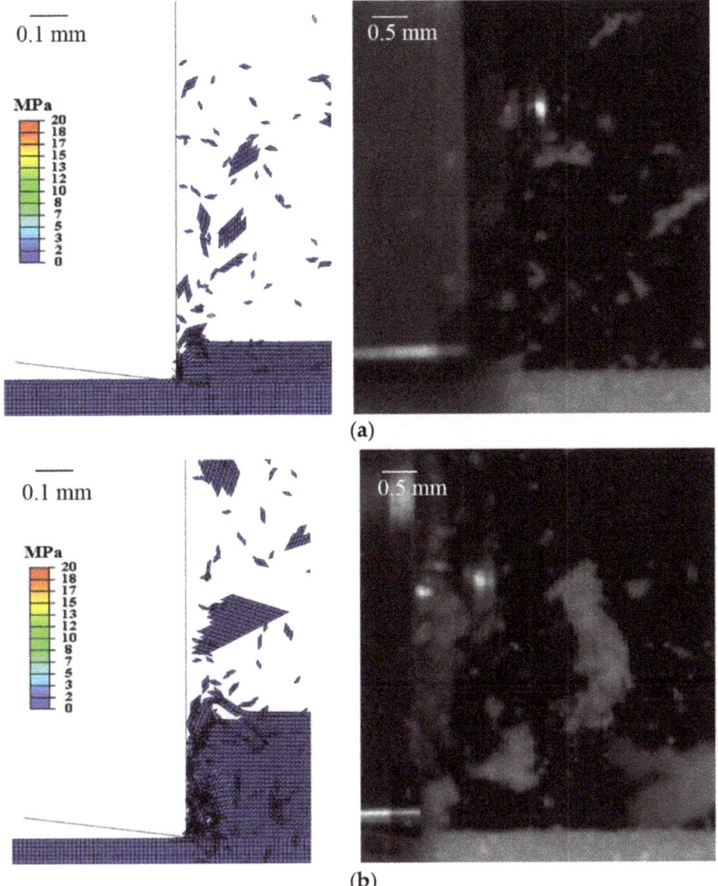

Figure 7. Simulated and experimentally measured chip formation of the 30 pcf with (**a**) depth of cut (DOC) = 0.1 mm and (**b**) DOC = 0.3 mm.

For the 40 pcf, the same scaling factor of 0.02 is used, which corresponds to a maximum of 0.00294 mm deformation (29.4% elongation). This value also makes the workpiece more ductile than the 30 pcf (12.8% elongation). The simulation result of 40 pcf at 0.1 mm DOC and corresponding experimental observations are shown in Figure 8a. Different from 30 pcf at 0.1 mm DOC, bigger and similarly-sized chips are generated with dusty debris around. This phenomenon also indicates a more ductile behavior as tested in Figure 4.

When the same scaling factor is applied to the case of 0.3 mm DOC, the simulation of the cutting process starts to show unstable chip formation, as shown in Figure 8b,c at different time steps. Cracks can propagate ahead of the cutting tool motion to generate large chips and sudden fracture along the cutting direction to shear the chip layer. This phenomenon is also seen in the experiment, though the unstable cracks into the workpiece could not really be captured due to the material uncertainty and the randomness of cracks.

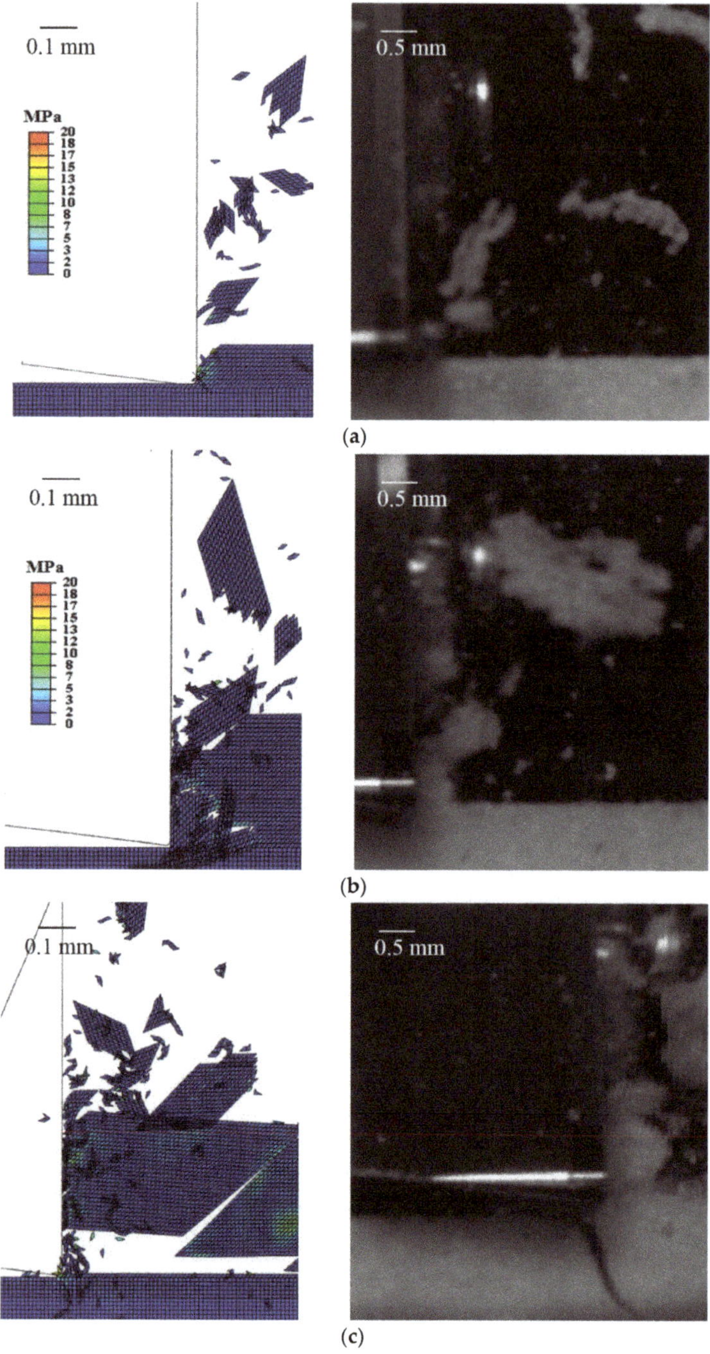

Figure 8. Simulated and experimentally measured chip formation of the 40 pcf with (**a**) DOC = 0.1 mm, (**b**) DOC = 0.3 mm, and (**c**) DOC = 0.3 mm at a later time step with a sudden crack propagation.

4.2. Cutting Force

Figure 9a shows the cutting forces measured from four repeated tests for 30 pcf at DOC = 0.1 mm. Force profiles are oscillating due to the brittle nature of the material. The system noise is assumed minimal considering the system rigidity. During a roughly 0.14 s cutting period, the cutting forces can reach and stay at a certain level, namely the steady cutting, and then drop toward the end. That said, the simulation length of about 0.01 s is enough to reach the steady cutting to extract the force. According to the scaling factor f = 0.02 used in these simulations, the simulated force is scaled by 50 times ($1/f$) and overlaid on Test 4, shown by the comparison in Figure 9b. Since the simulation ran at every 0.00006 s increment, the sampling frequency is equivalent to 16.7 kHz as opposed to 2 kHz of the experiment. The averaged force of simulation is 12.5 N, and the experimental average across the steady cutting is about 9 N. Although the forces are at a similar magnitude, the simulated force is oscillating much more significantly (0 to 35 N). These discrepancies may be attributed to the fact that embedded CZ elements have a different property from the main elements and less deformability. Such an oscillating profile is seen in all simulation cases of 30 and 40 pcf at 0.1 and 0.3 mm DOCs.

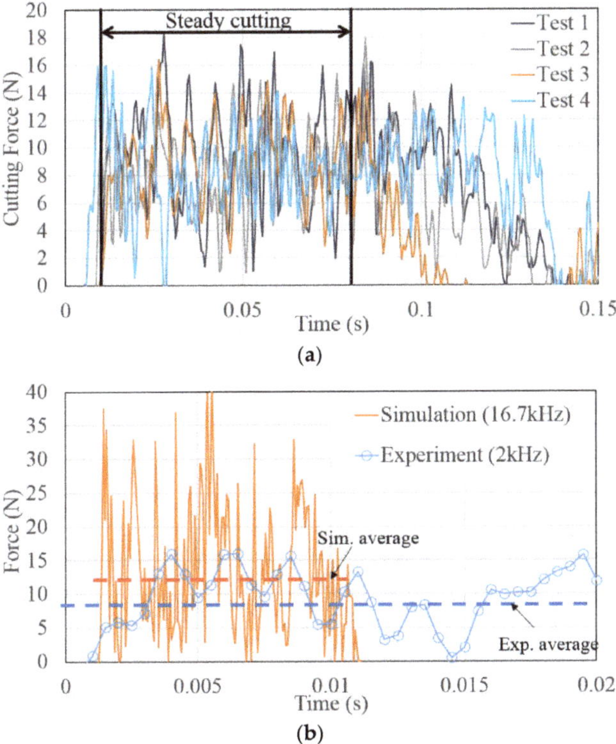

Figure 9. (**a**) Experimentally measured cutting forces of 30 pcf at DOC = 0.01 mm and (**b**) the comparison between the experiment and the simulated, scaled cutting force.

Figure 10 compares all simulated cases with the corresponding experiments in terms of the average force of cutting, where the error bars stand for one standard deviation from the four replicated tests. The overall trend of model prediction agrees with the experiments in different materials and depths of cut. However, the simulated forces are always higher by 30% to 50%, likely due to an over-estimated fracture energy or non-linearity of the cutting force to the cutting energy. The causes of oscillating and overestimated force will be elaborated more in the discussion section. Nonetheless, based on the results, the concept of ECZ–FEM is considered viable to approximate the magnitude of cutting force and to predict the changes of cutting force and chip behavior in different brittle cutting scenarios.

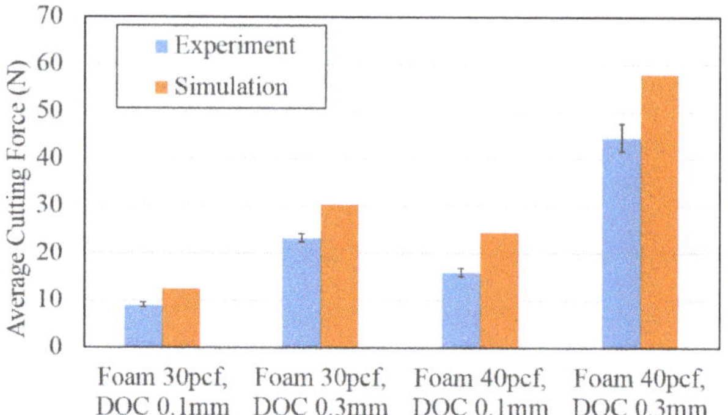

Figure 10. Comparisons between all simulated cutting forces and experimentally measured cutting forces (averaged).

5. Discussion

In ECZ–FEM, the key to a successful simulation is choosing an appropriate scaling factor by calibrating the model behavior with an experiment. As mentioned, CZ elements are determined by a traction–displacement relationship. When CZ elements are embedded in the workpiece, their allowable deflection can change the material ductility, and thus it must be limited. Figure 3 has shown how different scaling factors can change the chip formation from very ductile to brittle. Although limiting CZ deflection inevitably changes the material property (G_f), the effect on cutting force can be assumed linearly scaled under the assumption of 100% cutting energy conversion. This is reasonable because most of the brittle materials do not plastically deform and do not produce significant friction and frictional heat due to discontinuous chip formation.

The model predicts the relative behavior well among different materials and depths of cut, but the calculated cutting forces are always higher. One explanation is that it is due to the oscillating force profile, but it can also be caused by an overestimated fracture energy. The over-estimation can be from the difference between the static and dynamic fracture toughness, K_c. The fracture energy G_f is determined by the material toughness K_c, which is measured from a quasi-static test. Thus, the obtained K_c is the static fracture toughness while the actual dynamic toughness may be much lower, as reported in the literature [20,21]. However, it is technically challenging to measure a dynamic toughness at a comparable speed of cutting (10 m/min or 167 mm/s).

Another issue is the significant oscillating force profile as shown in Figure 9. This is because the model consists of embedded CZ elements which have different material properties and fewer degrees of freedom than those of the main elements. Therefore, the force can change drastically when the cutter makes a pass and the workpiece experiences deformation and damage. Another reason could be a non-self-contact definition of the main elements. This may result in intermittent contact between the tool and material and thus significant force changes. A much finer mesh with full contact definition can mitigate the problem at the cost of computational time.

6. Conclusions

This paper presents a fracture-based model for brittle material cutting using cohesive zone concept, namely ECZ–FEM. In this model, cohesive zone elements are embedded in the material body to allow free development of cracks to emulate the undetermined fracture during a cutting process. The research results have shown a certain degree of agreement with the experiment in terms of chip formation and cutting forces while also revealed some limitations. First, controlling the maximum deflection of the cohesive zone element through a scaling factor is a critical step in this method, and for that, an experimental calibration is necessary. This factor is currently determined on a qualitative basis in terms of chip size and crack propagation, because it is a behavior indicator instead of a property. Also, the current model is limited to brittle materials in order to scale the force linearly with the fracture energy. The model should also not be used for flexible material because the CZ mesh does not have enough degrees of freedom to handle deformation. For future work, modifications in CZ element or a new type of CZ element that can address these issues can further improve the model.

Author Contributions: B.T. developed the proposed model, conducted and analyzed the experiment, and wrote the manuscript. B.L.T. conceived the model concept, provided general guidance to this research, wrote and edited the final manuscript.

Funding: This research was partially funded by the Office of Energy Efficiency and Renewable Energy, U.S. Department of Energy, grant number DE-EE0008605.

Acknowledgments: The authors acknowledge the support from Texas A&M University and Texas A&M Engineering Experiment Station (TEES).

Conflicts of Interest: The authors declare no conflict of interest.

References

1. Liu, D.F.; Cong, W.L.; Pei, Z.J.; Tang, Y.J. A cutting force model for rotary ultrasonic machining of brittle materials. *Int. J. Mach. Tools Manuf.* **2012**, *52*, 77–84. [CrossRef]
2. Rao, G.V.G.; Mahajan, P.; Bhatnagar, N. Micro-mechanical modeling of machining of FRP composites–Cutting force analysis. *Compos. Sci. Technol.* **2007**, *67*, 579–593. [CrossRef]
3. Umer, U.; Ashfaq, M.; Qudeiri, J.; Hussein, H.; Danish, S.; Al-Ahmari, A. Modeling machining of particle-reinforced aluminum-based metal matrix composites using cohesive zone elements. *Int. J. Adv. Manuf. Technol.* **2015**, *78*, 1171–1179. [CrossRef]
4. Santiuste, C.; Soldani, X.; Miguélez, M.H. Machining FEM model of long fiber composites for aeronautical components. *Compos. Struct.* **2010**, *92*, 691–698. [CrossRef]
5. Usui, S.; Wadell, J.; Marusich, T. Finite element modeling of carbon fiber composite orthogonal cutting and drilling. *Procedia CIRP* **2014**, *14*, 211–216. [CrossRef]
6. Yan, X.; Reiner, J.; Bacca, M.; Altintas, Y.; Vaziri, R. A study of energy dissipating mechanisms in orthogonal cutting of UD-CFRP composites. *Compos. Struct.* **2019**, *220*, 460–472. [CrossRef]
7. Umbrello, D.; M'saoubi, R.; Outeiro, J. The influence of Johnson–Cook material constants on finite element simulation of machining of AISI 316L steel. *Int. J. Mach. Tools Manuf.* **2007**, *47*, 462–470. [CrossRef]
8. Shrot, A.; Bäker, M. Determination of Johnson–Cook parameters from machining simulations. *Comput. Mater. Sci.* **2012**, *52*, 298–304. [CrossRef]
9. Shi, J.; Liu, C.R. The influence of material models on finite element simulation of machining. *J. Manuf. Sci. Eng.* **2004**, *126*, 849–857. [CrossRef]
10. Takabi, B.; Tai, B.L. A review of cutting mechanics and modeling techniques for biological materials. *Med. Eng. Phys.* **2017**, *45*, 1–14. [CrossRef] [PubMed]
11. Takabi, B.; Tajdari, M.; Tai, B.L. Numerical study of smoothed particle hydrodynamics method in orthogonal cutting simulations–Effects of damage criteria and particle density. *J. Manuf. Processes* **2017**, *30*, 523–531. [CrossRef]
12. Turon, A.; Davila, C.G.; Camanho, P.P.; Costa, J. An engineering solution for mesh size effects in the simulation of delamination using cohesive zone models. *Eng. Fract. Mech.* **2007**, *74*, 1665–1682. [CrossRef]
13. Dong, X.; Shin, Y.C. Multi-scale genome modeling for predicting fracture strength of silicon carbide ceramics. *Comput. Mater. Sci.* **2018**, *141*, 10–18. [CrossRef]
14. Paulino, G.; Zhang, Z. Cohesive modeling of propagating cracks in homogeneous and functionally graded composites. In Proceedings of the 5th GRACM International Congress on Computational Mechanics, Limassol, Cyprus, 29 June–1 July 2005.
15. Liang, S.Y.; Shih, A.J. *Analysis of Machining and Machine Tools*; Springer: Boston, MA, USA, 2016.
16. Liu, J.; Bai, Y.; Xu, C. Evaluation of ductile fracture models in finite element simulation of metal cutting processes. *J. Manuf. Sci. Eng.* **2014**, *136*, 011010. [CrossRef]

17. Espinosa, H.D.; Zavattieri, P.D. A grain level model for the study of failure initiation and evolution in polycrystalline brittle materials. Part II: Numerical examples. *Mech. Mater.* **2003**, *35*, 365–394. [CrossRef]
18. Feng, J.; Chen, P.; Ni, J. Prediction of surface generation in microgrinding of ceramic materials by coupled trajectory and finite element analysis. *Finite Elem. Anal. Des.* **2012**, *57*, 67–80. [CrossRef]
19. Sawbones Inc., General Catalog. Available online: https://www.sawbones.com/wp/wp-content/uploads/2017/07/Gen-393Catalog-ReVamp-V1.pdf (accessed on 13 March 2019).
20. Kobayashi, A.; Mall, S. Dynamic fracture toughness of Homalite-100. *Exp. Mech.* **1978**, *18*, 11–18. [CrossRef]
21. Kobayashi, T.; Yamamoto, I.; Niinomi, M. Introduction of a new dynamic fracture toughness evaluation system. *J. Test. Eval.* **1993**, *21*, 145–153.

Journal of
*Manufacturing and
Materials Processing*

Article

Five-Axis Machine Tool Coordinate Metrology Evaluation Using the Ball Dome Artefact Before and After Machine Calibration

Heidarali Hashemiboroujeni *, Sareh Esmaeili Marzdashti, Kanglin Xing and J.R.R. Mayer

Advanced Manufacturing Department, Faculty of Mechanical Engineering, Polytechnique Montréal, University of Montreal, Montreal, QC H3T 1J4, Canada; sareh.esmaeili-marzdashti@polymtl.ca (S.E.M.); kanglin.xing@polymtl.ca (K.X.); rene.mayer@polymtl.ca (J.R.R.M.)
* Correspondence: ali.hashemi@polymtl.ca; Tel.: +1-438-993-6750

Received: 31 December 2018; Accepted: 1 February 2019; Published: 3 February 2019

Abstract: Now equipped with touch trigger probes machine tools are increasingly used to measure workpieces for various tasks such as rapid setup, compensation of final tool paths to correct part deflections and even verify conformity to finished tolerances. On five-axis machine tools, the use of data acquired for different rotary axes positions angles brings additional errors into play, thus increasing the measurement errors. The estimation of the machine geometric error sources, using such methods as the scale and master ball artefact (SAMBA) method, and their use to calibrate machine tools may enhance five-axis on-machine metrology. The paper presents the use of the ball dome artefact to validate the accuracy improvement when using a calibrated model to process the machine tool axis readings. The inter-axis errors and the scale gain errors were targeted for correction as well the measuring tool length and lateral offsets. Worst case and mean deviations between the reference artefact geometry and the on-machine tool measurement is reduced from 176 and 70 μm down to 31 and 12 μm for the nominal and calibrated machine stylus tip offsets respectively.

Keywords: coordinate metrology; on-machine measurement; ball dome artefact; calibration; machine tool

---

1. Introduction

Machine tools with three, five or more axes are now equipped with touch trigger probes to accomplish metrology tasks such as tool path re-planning [1] or setup location and finishing path correction for the workpieces and even to evaluate the conformity of the finished machined parts [2]. The machine tool accuracy directly affects its ability to be used for such tasks. Accuracy is defined in the VIM (International Vocabulary of Metrology) as "closeness of agreement between a measured quantity value and a true quantity value of a measurand" [3] thus that it includes both systematic and so-called non-repeatable effects. A similar approach is used in the ISO (International Organization for Standardization) standard on machine tool accuracy [4].

On coordinate measuring machines, a probe head with two rotary axes is used to gain access to features on complex parts. The resulting change in the position of the stylus tip with respect to the machine foundation is handled by calibrating the change in this position through the probing of a reference ball at a fixed position in the machine base frame. A similar approach could be adopted on a five-axis machine, but it is not probably due to the limited available space on the workpiece table. Instead, the approach here is to rely on the measured angular positions of the rotary axes to perform the computation of the stylus tip in the workpiece table frame. Performing such calculation using a nominal, error free machine model will likely result in coordinates of a similar level of accuracy as the machine tool itself. Improvement in the computed coordinates, as was done on coordinate measuring machines [5,6], is possible through the use of a rigid body kinematic model incorporating known machine errors. Using the mathematical models to simulate the machine tool geometry is the main concept to compensate the error parameters.

A variety of approaches have been proposed to acquire the machine error parameters [7,8], using touch trigger probes [9], scanning probes, ball bars [10], laser interferometers, and laser trackers are some methods which have been applied to this task. A pseudo 3D grid configured from a kinematically relocated calibrated 2D ball plate [11] was proposed for testing and calibrating machine tools but it was used for a 3-axis vertical machine. By increasing the number of machine axes, with rotary axes, the machine geometry becomes more complex and the number of error sources increases. Assessing the out of sphericity by probing 25 points on a precise ball mounted on the machine tool table, for various rotary axes indexations was used to assess the coordinate measurement accuracy of a five-axes machine tool before and after considering the machine's error parameters [12]. However, no traceability is provided to the meter. The ball dome artefact, proposed by Mayer and Hashemi [13] is made of Invar, to eliminate the thermal effects deformation, was developed to estimate a five-axis machine tool metrology performance. Calibrating the coordinates of the balls to obtain reference values provides this traceability.

Machining a part and then measuring it by a coordinate measuring machine (CMM) is a common industrial method to check the accuracy of a machine but it is not only an expensive and time consuming method but also it is just applicable to the machining mode and it is not useful for machine evaluation in the coordinate measuring mode [14,15].

In this paper, an alternative calibration verification method is defined for a five-axis machine center when all five axes contribute to the measurement. First, the SAMBA method for machine calibration [16] is briefly explained. Then the mathematical model used to compensate the machine readings using its topology and error parameters is presented. It is followed by the SAMBA experimental probing procedure, which produces the calibrated machine stylus tip offsets. Finally, the newly designed ball dome artefact is used to validate the SAMBA calibrated model for a five-axis machine tool used as a five-axis coordinate measuring machine.

2. SAMBA Calibration Method

The machine tool error parameters are gathered using the scale and master balls artefact (SAMBA) method, which consists in probing special artefacts and using the raw probing data to estimate the machine error parameters as an indirect method through a mathematical model. The SAMBA hardware part is composed of a reconfigurable uncalibrated master ball artefact (RUMBA) and a length standard; all mounted on the machine table. The processing of the raw probing data allows estimating the machine errors parameters, the artefact positions, the stylus tip coordinates (as the tool), and the volumetric errors.

Let the topology of the machine be wCBXFZYSt wherein the workpiece branch includes C-, B- and X-axis and the tool branch includes the Z- and Y-axis and the spindle. The two branches are linked by the foundation frame F. W, S and t stand for the workpiece, the spindle and the tool, respectively. The nominal kinematics of the machine is

$$^{w_n}T_{t_n} = \left(^F T_X{}^X T_B{}^B T_C{}^C T_{w_n}\right)^{-1} \left(^F T_Z{}^Z T_Y{}^Y T_S{}^S T_{t_n}\right) \tag{1}$$

where the first parenthesis is the homogeneous transformation matrix (HTM) of the workpiece to the frame and the second one is the HTM of the tool to the frame. However, the kinematics of a real machine contains the errors as follows:

$$^{w_a}T_{t_a} = \left(^F T_{X_0}{}^{X_0} T_{X'_0}{}^{X'_0} T_X{}^X T_{X'}{}^{X'} T_{B_0}{}^{B_0} T_{B'_0}{}^{B'_0} T_B{}^B T_{B'}{}^{B'} T_{C_0}{}^{C_0} T_{C'_0}{}^{C'_0} T_C{}^C T_{C'}{}^{C'} T_{w_n}{}^{w_n} T_{w_a}\right)^{-1}$$
$$\left(^F T_{Z_0}{}^{Z_0} T_{Z'_0}{}^{Z'_0} T_Z{}^Z T_{Z'}{}^{Z'} T_{Y_0}{}^{Y_0} T_{Y'_0}{}^{Y'_0} T_Y{}^Y T_{Y'}{}^{Y'} T_S{}^S T_{S'}{}^{S'} T_{t_n}{}^{t_n} T_{t_a}\right) \tag{2}$$

where X_0, Y_0, Z_0, B_0, C_0, w_n, S, and t_n are the nominal joint positions. X'_0, Y'_0, Z'_0, B'_0, and, C'_0 are the actual joint positions before movement. X, Y, Z, B, and C describe the nominal motion and X', Y', Z', B', C', w_a, S', and t_a describe the error motions. The erroneous five-axis machine requires 30 intra-axis errors (error motions) and eight inter-axis errors (axis location errors). Considering the two spindle lateral offsets and the three linear axis scale gain errors add another five parameters. However, different error models have been studied which typically contain all or a few of those errors [16,17]. The "13" machine error model, describing the erroneous machine, is studied in this paper due to its advantages such as short measuring time and simple indexation design. This error model consists of eight axis location errors, the two spindle offsets, and the three linear gains errors [16]. The model mainly contains inter-axis errors. However, the three linear gain errors, EXX1, EYY1 and EXX1 associated with the intra-axis errors EXX, EYY, and EZZ, respectively, are significant error sources and thus they are added to the model.

The strategy of SAMBA method is applied wherein B- and C-axis fully rotate. By releasing ball positions, which are not accessible, by the touch probe in some indexations, a number of joint positions are achieved for which the following Jacobian is constructed

$$E_V = JE_P \tag{3}$$

where E_V is the volumetric error, J is the Jacobian and E_P contains the machine error parameters. Provided a well-conditioned system a least square solution is found via the pseudo inverse

$$E_P = J^\dagger E_V \tag{4}$$

where $J^\dagger$ is the pseudo-inverse of J. The main steps of the SAMBA method are as follows: machine error model selection, artefact selection, indexation design (relative positions of the rotary axes) and verification, probing G-code generation, probing on the real machine tool and data processing.

3. SAMBA Test on Experimental Machine Tool

Figure 1 shows the probing process with an MP700 Renishaw touch trigger probe of the SAMBA method of four accessible master balls and one scale bar artefact installed on the pallet of the HU40-T machine tool and Table 1 presents the nominal position of the ball centers. The tough trigger probe contacts the workpiece, which triggers the acquisition of the X-, Y- and Z- axis readings. The master ball is always measured twice. For each of these measurements, the master ball artefact is measured with fast and slow probing speed. The first measurement is to get a better estimate of the ball position, before re-measuring it using this new center as a target for the probing. For the second measurement, the probing approaches are adjusted to ensure the spherical surface is touched with an approach close to the local surface normal. The second master ball artefact positions are recorded.

For each measurement, the same probing strategy is applied. It is probed at +45 and −135 degree, to get a plane for the measurements at –45 and +135 degree. Then a point on the pole is taken. A simple geometric calculation is used to estimate the centre. For the SAMBA probing process, the Renishaw MP700 touch trigger probe has negligible pre-travel variation errors (0.25 um) thus it is not compensated.

A total of 109 balls probing for 32 angular axes indexations pairs are recorded from which 13 machine error parameters, six balls coordinates and three tool coordinates are estimated. Axes indexations include $0°$, $\pm 10°$, $\pm 30°$, $\pm 60°$, and $\pm 90°$ for the B-axis; from zero to $360°$ and reverse for the C-axis by $90°$ steps and from zero to $360°$ for the spindle axis by $90°$ steps. During the measurement, the laboratory temperature varied between 21 and 23 $°C$ and the machine tool is started in the cold condition before each measurement. The test was repeated 11 times on different days over a two month period.

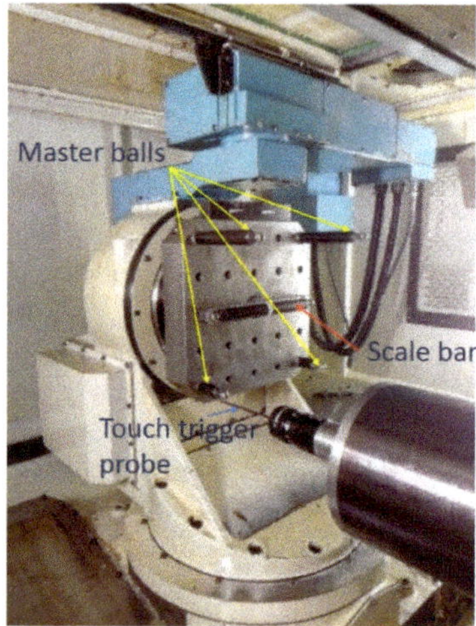

Figure 1. Scale and master ball artefact (SAMBA) measurement process on the HU40-T five-axis machine tool.

Table 1. SAMBA balls position.

Ball Identifier	X (mm)	Y (mm)	Z (mm)
1	−152.4000	0	40.8550
2	152.4000	0	40.4350
3	160.0000	160.0000	177.8350
4	−160.0000	160.0000	177.6950
5	−160.0000	−160.0000	75.6050
6	160.0000	−160.0000	76.0450

4. Machine Tool Estimated Error Parameters

The estimated machine tool error parameters obtained from the SAMBA method are listed in Table 2. Figure 2 illustrates the error terms in the machine kinematic chain.

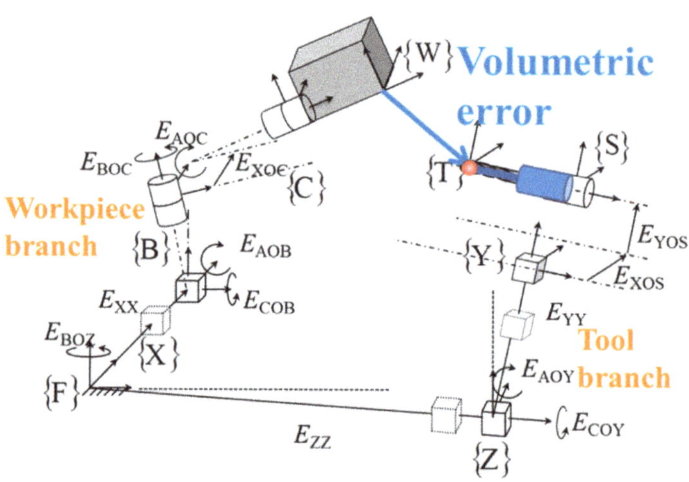

Figure 2. Estimated error parameters in the machine kinematic chain (modified from reference [16]).

Table 2. Estimated error parameters for the 13-error model (nomenclature as per ISO230-1:2012).

ISO Name	Parameter Description	Value
E_{A0B}	Out-of-squareness angle of the B-axis relative to the Z-axis	−11 µrad
E_{C0B}	Out-of-squareness angle of the B-axis relative to the X-axis	−8 µrad
E_{X0C}	Offsets between the B and C axes	−0.105 mm
E_{A0C}	Out-of-squareness of the C-axis relative to the B-axis	−11 µrad
E_{B0C}	Out-of-squareness of the C-axis relative to the X-axis	−8 µrad
E_{B0Z}	Out-of-squareness of the Z-axis relative to the X-axis	−13 µrad
E_{A0Y}	Out-of-squareness of the Y-axis relative to the Z-axis	−18 µrad
E_{C0Y}	Out-of-squareness of the Y-axis relative to the X-axis	21 µrad
E_{Y0S}	Offset of the spindle relative to the C-axis in Y	0.020 mm
E_{X0S}	Offset of the spindle relative to the B-axis in X	−0.106 mm
E_{XX1}	Positioning linear error of the X-axis	−16 µm/m
E_{YY1}	Positioning linear error of the Y-axis	11 µm/m
E_{ZZ1}	Positioning linear error of the Z-axis	21 µm/m

5. The Ball Dome Artefact

In order to evaluate the machine metrology performance across the entire machine workspace, the maximum number of artefact balls should be accessible for probing for a broad range of angular axis positions. The ball dome artefact structure includes three semi-circular arcs attached together at their mid-point, with both ends fixed to a base ring. The result is a quasi-hemispherical structure holding 25 balls. In addition, there are three balls on the base ring and four balls on the base plate, which provide stable points to define a reference coordinate system. This design allows testing the machine for the full range of rotary axis motion. On this machine tool, the B-axis and C-axis rotation range are –90° to +90° and 0° to 360° respectively. The ball dome artefact is shown in Figure 3. To limit thermal effects the artefact structure is made of Invar. The measurement repeatability is affected by the clamping force that is applied to hold the artefact on the base plate, and by elastic deformation caused by a changing gravity vector. The reported measurement repeatability for clamping and gravity deflection was on average of the order of 0.6 and 6.5 µm respectively [13]. The measurement uncertainty for the artefact ball center once mounted on the machine tool is also reported at 5.3 µm (k = 2) [13].

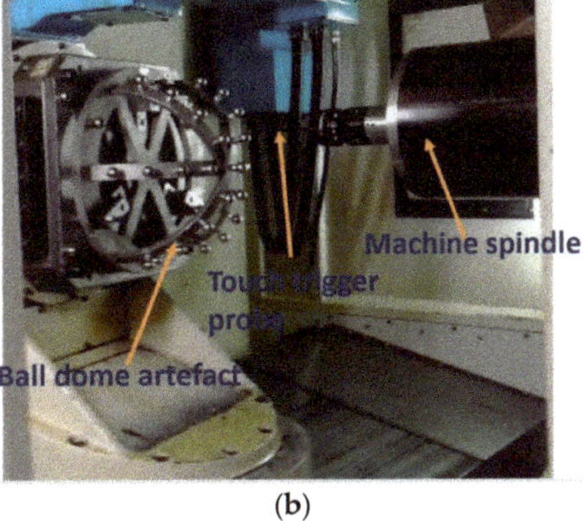

(a) (b)

Figure 3. (a) Ball dome artefact, (b) the ball dome holds on the machine tool work table.

6. Ball Dome Probing

The ball dome artefact is used to evaluate the machine tool metrology performance with and without the machine calibration. The artefact balls coordinates were measured to an uncertainty of 5.3 µm; Table 3 lists the measured coordinates. The artefact then is probed on the machine tool. The machine measured ball center coordinates in the machine table frame are calculated using the axis position readings and either the nominal model or the calibrated one, Equation (1). Then they are compared to the reference artefact coordinates. Because coordinate metrology generally requires accessing some features from different angles, the artefact should be probed at various machine indexations to ensure that different rotary axes positions are involved in the measurement process. The ball dome artefact is probed in 24 different machine rotary axes indexations, from $-90°$ to $+90°$ for the B axis and from $0°$ to $360°$ for the C axis. At each indexation, the maximum numbers of balls, which are accessible for the probing tool, are measured. A total of 613 ball centers were measured in about 15 h. All the other accessible balls centers are measured once at each pair of rotary axes indexation,

Table 3. Balls coordinates measured on the coordinate measuring machine (CMM) and used as calibrated values.

Ball	X (mm)	Y (mm)	Z (mm)
1	−0.2132	−165.1204	51.5225
2	−0.2629	−139.6651	99.1026
3	−0.3403	−101.2336	136.8692
4	−0.3955	−52.6576	160.2703
5	−0.2887	52.9376	159.8028
6	−0.2151	100.6740	136.3087
7	−0.2995	139.3806	98.7283
8	−0.0574	164.8339	51.0565
9	−151.3660	−87.3181	54.0706
10	−128.5393	−73.7984	104.8753
11	−92.6593	−53.1975	144.6691
12	−48.5258	−27.6129	169.9002
13	48.2969	28.3799	170.6793
14	92.7368	53.9159	145.7798
15	128.8048	74.6037	105.5009
16	151.8661	87.8615	54.1932
17	−158.5380	91.4808	57.2096
18	−134.0660	77.0156	110.0194
19	−97.3725	55.6279	151.3944
20	−51.4891	29.2089	178.4874
21	−0.2529	−0.3028	188.9155
22	50.7425	−29.6543	180.3483
23	97.7621	−56.4384	153.8900
24	135.6217	−78.1585	111.5776
25	159.6956	−92.6723	57.6537
26	80.9629	−140.3990	9.8332
27	−162.7489	0.3372	9.3833
28	80.9960	141.0359	9.9255
29	124.7426	−174.8164	−27.5606
30	−125.2386	−174.7670	−27.5602
31	−124.9456	175.3488	−27.5602
32	124.7427	175.3130	−27.5607

7. Stylus Tip Offsets Calculation

On a five-axis machine tool, there will be a situation when measurements are taken at different rotary axes positions are combined to analyse particular geometric characteristics of the workpiece. In such cases, the stylus tip coordinates are needed. These coordinates can be obtained using different approaches yielding different quality of results. In addition, the machine's own geometry, as for a coordinate measuring machine, needs to be calibrated and compensated. However, most machine tools are not geometrically calibrated. In this section, various ways to calibrate the stylus tip offsets and the machine geometry are presented. The ball dome is then used as a reference to evaluate the effectiveness of the calibrated models. One of the parameters studied is the effect of the stylus tip offsets. The term stylus tip offsets here stand for the coordinates of the stylus tip center of the touch trigger probe relative to last machine tool branch axis frame, in our case the Y-axis frame. The stylus tip offsets can either be the nominal value for the tool length or values estimated through the SAMBA algorithms by probing one or more balls at various positions of the machine rotary axes. Table 4 lists the various stylus tip offsets used and how they are obtained. The ball dome data was processed either using a nominal machine model with null error parameters or using error parameters estimated from the SAMBA method.

N1-Nominal machine model, nominal tool (tool item N1):

The nominal tool is assumed to have zero lateral offsets and the tool length, as measured by the machinist during tool setting, as a negative z value.

N2-Nominal machine, estimated tool from a single ball dome ball (tool item N2):

The other approach to determine the stylus tip offsets is to use a nominal machine to estimate the stylus tip offsets. The tool length ($-z$ value) and lateral offsets in x and y are estimated by using a single ball on the ball dome, which is located close to the ring section; no machine error parameter is estimated, and the parameters are set to zero. The ball and the tool coordinates are the only estimated variables to explain the machine readings.

N3-Nominal machine:

A similar process as for N2 but using all ball dome balls at once.

S1-SAMBA estimated machine, the tool from machinist for ball dome probing (tool item S1):

The same tool as for item N1, reported by the machinist, is used in this case. However, the ball dome coordinates are calculated based on the machine estimated from the SAMBA process, from one year ago.

S2-SAMBA estimated machine, an estimated tool from a single ball dome ball (tool item S2):

The tool x, y and z coordinates are estimated in order to best explain the machine readings while using the machine error parameters estimated by the SAMBA process from one year ago. The calibrated ball dome coordinates are not used.

S3-SAMBA estimated machine, an estimated tool from all ball dome balls (tool item S3):

As for S2 but all the ball dome balls are used for the tool estimation.

S4-SAMBA estimated machine, manually estimated tool (tool item S4):

The tool is estimated during the machine calibration using the SAMBA method. However, the stylus tip used to measure the ball dome was different from that used for the SAMBA calibration. In addition, during the dome measurement, the spindle was not rotated so that the tool could not be estimated independently from the spindle position. The spindle location was estimated during the SAMBA calibration conducted a year earlier. A complete machine, tool and ball coordinates estimation are conducted to explain the machine probing readings but only the tool coordinates are further used here. Vector subtraction is used to extract the tool geometry from the two vectors as illustrated in Figure 4 resulting in the following equation,

$$tool^a = tool^{ball-dome} - spindle^{SAMBA} \tag{5}$$

$$tool^a = \left(tool^n + \delta tool^{ball-dome} \right) - spindle^{SAMBA} \tag{6}$$

where *tool^n* is the nominal tool geometry used during ball-dome measurement and $\delta tool^{ball-dome}$ is the deviation of the tool geometry calculated by Equation (4) and

$$\delta spindle^{SAMBA} = [E_{XOS} \quad E_{YOS} \quad 0] \tag{7}$$

where E_{XOS} and E_{YOS} are the two spindle lateral offset errors obtained by SAMBA method defined in Table 2. The kinematics of the machine tool accompanied by the two offset errors contributing in the tool twist estimation are illustrated in Figure 2.

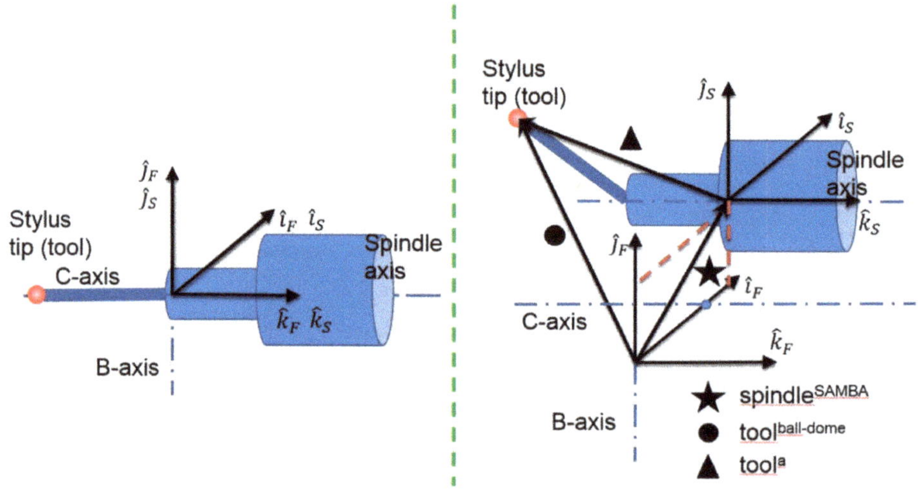

Figure 4. Left: For a nominal machine, the reference frame of the spindle coincides with the B- and C-axis crossing point, which is also the machine foundation frame. For case N1 the tool only has a non-zero z coordinate. **Right**: The non-nominal machine estimated by the SAMBA method has spindle offsets in x from the B-axis and in y from the C-axis. This right-side diagram also illustrates the cases of an imperfect tool with lateral offsets in x and y.

Table 4. Various stylus tip offsets.

Item	Tool Description	Stylus Tip Offsets (mm)		
		tX	tY	tZ
N1	Nominal machine with tool from machinist for ball dome probing	0	0	−326.717
N2	Nominal machine with estimated tool by using one ball on ball dome	−0.105	0.001	−326.725
N3	Nominal machine with estimated tool by using all balls on ball dome	−0.046	−0.002	−326.735
S1	Estimated machine by SAMBA, tool from machinist for ball dome probing	0	0	−326.717
S2	Estimated machine by SAMBA, estimated tool from one ball	−0.011	0.002	−326.749
S3	Estimated machine by SAMBA, estimated tool from all the artefact balls	−0.012	−0.012	−326.746
S4	Estimated machine by SAMBA, tool from full estimation with dome, manually calculated tool	−0.003	−0.014	326.754

8. Deviation Results

Table 5 shows the maximum and average deviation calculated for different models. The deviation is between artefact reference probing coordinate from CMM measurements and artefact probing coordinate from the machine tool as a coordinate measuring system. A least square fitting algorithm is applied to best match the two sets of coordinates. For every single ball, the coordinate deviations in x, y and z and the deviation norm, R, are calculated. The maximum and average values of R considering all ball dome balls are calculated.

Table 5. Maximum and mean deviation for different models (μm).

Item	Tool	Maximum Deviation μm	Average Deviation μm
N1	Nominal machine with tool from machinist for ball dome probing	176	70
N2	Nominal machine with estimated tool by using one ball on ball dome	138	88
N3	Nominal machine with estimated tool by using all balls on ball dome	140	60
S1	Estimated machine by SAMBA, tool from machinist for ball dome probing	56	30
S2	Estimated machine by SAMBA, estimated tool from one ball	34	19
S3	Estimated machine by SAMBA, estimated tool from all the artefact balls	32	12
S4	Estimated machine by SAMBA, tool from full estimation with dome, manually calculated tool	31	16

Figure 5 shows the deviation between the calibrated artefact and machine tool measured coordinates while using SAMBA estimated machine and manually estimated stylus tip offsets (item S4), which are used for comparison. The vectors (arrows) are the 3D deviation for every single ball while the reference values are the artefact reference coordinate measured on a CMM and each color represents a specific machine tool axes indexation out of the 24 indexations. Each vector has three Cartesian components; the length of each vector is calculated by Equation (8):

$$dR = \sqrt{\left(dx^2 + dy^2 + dz^2\right)} \tag{8}$$

The maximum and average deviations (vector lengths) for the 25 balls at 24 indexations are presented in Table 4.

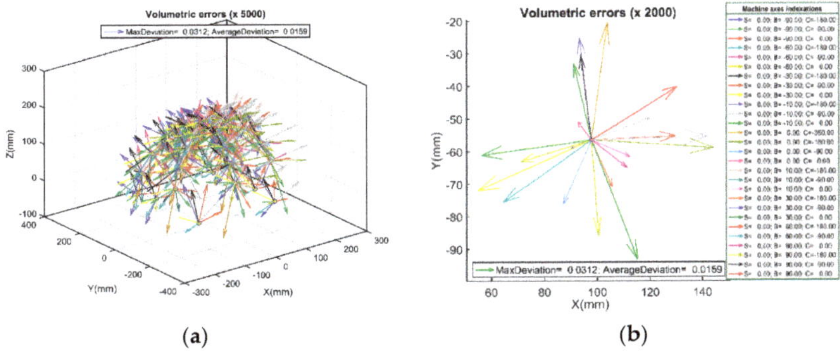

Figure 5. Plotted deviation between compensated artefact by SAMBA and calibrated artefact, manually estimated tool is applied (item S4); (**a**) 3D view; the legend of the arrows' colors is as on figure b. (**b**) Deviation for one ball in X-Y view for different machine axes and indexations (units are millimeter).

Figure 6 presents the deviation while using the stylus tip offsets calculated based on just one ball on the artefact (item S2). In this case, to lighten the plots, the deviation arrows just for seven selected balls are shown through the artefact.

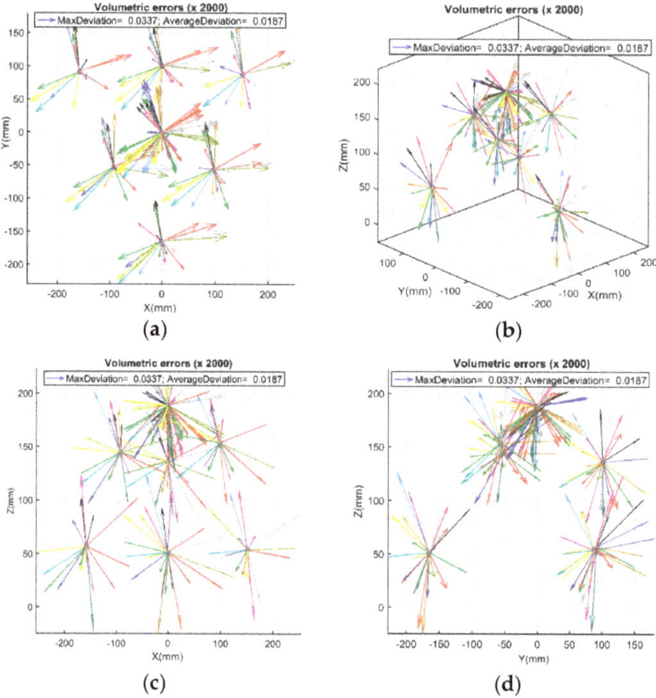

Figure 6. Plotted deviation between compensated artefact by SAMBA while considering one ball to model the tool and calibrated artefact for seven balls, estimated tool by probing one ball (item S2) (units are millimetre). (**a**) Deviation arrows, X-Y view, (**b**) 3-D view, (**c**) X-Z view, (**d**) Y-Z view

On the other hand, as it mentioned in Table 5, while the non-calibrated machine and nominal stylus tip offsets are used, the maximum and average deviations are 176 and 70 μm respectively.

9. Discussion

The SAMBA calibrated machine tool errors parameters are used to compensate the machine for the purpose of on-machine coordinate metrology. The considered errors are eight axis location errors, two spindle lateral offsets, three linear axis positioning scale gain errors and the stylus tip center coordinates (the tool) relative to the spindle frame. The ball dome artefact is used to evaluate the accuracy of the compensated machine. The ball dome includes 25 balls on a quasi-hemispherical envelop fabricated of Invar, which is clamped on kinematic supports to reduce clamping distortion.

The machine measuring performance when no calibration is applied neither for the machine geometry nor for the stylus tip offsets, displays the maximum and average deviations equal to 176 and 70 μm respectively. Calibrating the machine geometry based on the SAMBA estimated error parameters improves the machine performance and reduces the maximum and average deviation to 56 and 30 μm, respectively, a 60% improvement. Another important error contributor is the stylus tip offsets. There are two options to estimate the stylus tip offsets; the first one is using just one ball on the artefact which leads to 34 and 19 μm as the maximum and average deviations. The results achieved by using the second option, which stands on using all balls, are 32 and 12 μm. The other choice for the stylus tip offsets is achieved by vector calculation between the estimated tool from ball dome data only and estimated spindle from the SAMBA process. For this last case the maximum and average deviation are 31 and 16 μm respectively, the lowest maximum value obtained. The deviation reduction achieved by using calibrated machine and estimated stylus tip offsets is figured in Figure 7.

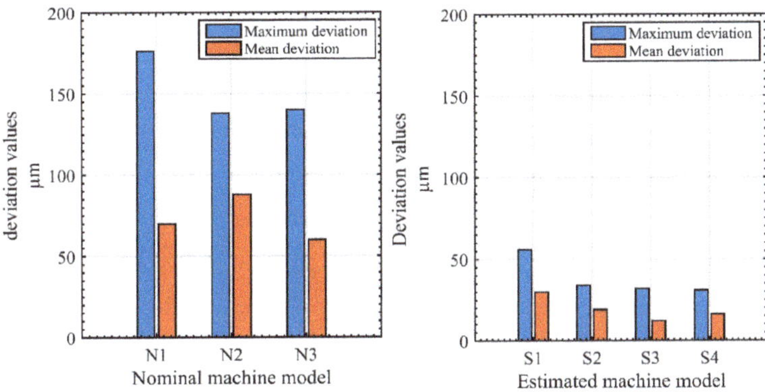

Figure 7. Deviation for nominal and estimated machine.

10. Conclusions

In this paper, the SAMBA method is used to calibrate the machine tool and to estimate the stylus tip offsets and then the efficiency of the calibration process and estimated models are verified by using the ball dome artefact. Due to the large size of the ball dome, during its measurement by the machine on various rotary axes indexations, all the linear and rotary axes motions are covered and then all the measurement results are transferred to the same reference frame. A single ball measurement would be the preferred option to estimate the stylus tip offsets. This approach provides the lowest average deviation.

Therefore, using the SAMBA calibration method accompanied with an optimized machine and stylus tip offsets has reduced the machine tool maximum and average volumetric errors by 82% and 83% respectively, while all the linear and rotary axes are involved in the coordinate measurement process.

Author Contributions: Conceptualization, J.R.R.M. and H.H.; methodology, H.H.; software, J.R.R.M.; validation, H.H.; formal analysis, H.H., S.E.M. and K.X.; investigation, H.H., S.E.M. and K.X.; resources, H.H. and K.X.; data curation, H.H. and S.E.M.; writing—original draft preparation, H.H.; writing—review and editing, J.R.R.M.; visualization, H.H., S.E.M.; supervision, J.R.R.M.; project administration, J.R.R.M.; funding acquisition, J.R.R.M.

Funding: This research was supported by the Natural Sciences and Engineering Research Council of Canada NSERC Canadian Network for Research and Innovation in Machining Technology–Phase 2: CANRIMT2.

Acknowledgments: The authors would like to thank the valuable support of CNC machine technicians, Guy Gironne and Vincent Mayer during the experimental tests.

Conflicts of Interest: The authors declare no conflict of interest.

References

1. Lasemi, A.; Xue, D.Y.; Gu, P.H. Tool path re-planning in free-form surface machining for compensation of process-related errors. *Int. J. Prod. Res.* **2014**, *52*, 5913–5931. [CrossRef]
2. Lee, W.C.; Wang, J.Y.; Wei, C.C. Improving Machining Accuracy by Automatic Compensation Based on the Off-line Measurement. In Proceedings of the 2017 IEEE International Conference on Consumer Electronics–Taiwan (ICCE-TW), Taipei, Taiwan, 12–14 June 2017.
3. JCGM 200. *International Vocabulary of Metrology—Basic and General Concepts and Associated Terms International Vocabulary of Metrology*; JCGM: Kasterlee, Belgium, 2012.
4. ISO. *Standard on Machine Tool Accuracy In 230-1*; ISO: Geneva, Switzerland, 2012.
5. Zhang, G.; Veale, R.; Charlton, T.; Borchardt, B.; Hocken, R. Error compensation of co-ordinate measuring machines. *CIRP Ann. Manuf. Technol.* **1985**, *34*, 445–448. [CrossRef]
6. Hocken, J.A.S.R.; Borchardt, B.; Lazar, J.; Reeve, C.; Stein, P. Three Dimensional Metrology. *CIRP Ann. Manuf. Technol.* **1977**, *26*, 403–408.

7. Schwenke, H.; Knapp, W.; Haitjema, H.; Weckenmann, A.; Schmitt, R.; Delbressine, F. Geometric error measurement and compensation of machines—An update. *CIRP Ann. Manuf. Technol.* **2008**, *57*, 660–675. [CrossRef]

8. Ibaraki, S.; Nagai, Y. Formulation of the influence of rotary axis geometric errors on five-axis on-machine optical scanning measurement-application to geometric error calibration by "chase-the-ball" test. *Int. J. Adv. Manuf. Technol.* **2017**, *92*, 4263–4273. [CrossRef]

9. Guiassa, R.; Mayer, J.R.R.; Kops, L. Predictive compliance based model for compensation in multi-pass milling by on-machine probing. *CIRP Ann. Manuf. Technol.* **2011**, *60*, 391–394. [CrossRef]

10. Lasemi, A.; Xue, D.Y.; Gu, P.H. Accurate identification and compensation of geometric errors of 5-axis CNC machine tools using double ball bar. *Meas. Sci. Technol.* **2016**, *27*, 055004. [CrossRef]

11. Bringmann, B.; Kung, A. A measuring artefact for true 3D machine testing and calibration. *CIRP Ann. Manuf. Technol.* **2005**, *54*, 471–474. [CrossRef]

12. Rahman, M.M.; Mayer, R.R.R. Performance of a five-axis machine tool as a coordinate measuring machine (CMM). *J. Adv. Mech. Des. Syst.* **2016**, *10*. [CrossRef]

13. Mayer, J.R.R.; Hashemiboroujeni, H. A ball dome artefact for coordinate metrology performance evaluation of a five axis machine tool. *CIRP Ann. Manuf. Technol.* **2017**, *66*, 479–482. [CrossRef]

14. Givi, M.; Mayer, J.R.R. Optimized volumetric error compensation for five-axis machine tools considering relevance and compensability. *CIRP J. Manuf. Sci. Technol.* **2016**, *12*, 44–55. [CrossRef]

15. Ibaraki, S.; Sawada, M.; Matsubara, A.; Matsushita, T. Machining tests to identify kinematic errors on five-axis machine tools. *Precis. Eng.* **2010**, *34*, 387–398. [CrossRef]

16. Mayer, J.R.R. Five-axis machine tool calibration by probing a scale enriched reconfigurable uncalibrated master balls artefact. *CIRP Ann. Manuf. Technol.* **2012**, *61*, 515–518. [CrossRef]

17. Mchichi, N.A.; Mayer, J.R.R. Axis location errors and error motions calibration for a five-axis machine tool using the SAMBA method. *Proc. CIRP* **2014**, *14*, 305–310. [CrossRef]

Journal of
Manufacturing and
Materials Processing

Article

Flexible Abrasive Tools for the Deburring and Finishing of Holes in Superalloys

Adrián Rodríguez, Asier Fernández, Luís Norberto López de Lacalle and
Leonardo Sastoque Pinilla *

Aeronautics Advanced Manufacturing Center, 48170 Zamudio, Spain; Adrian.Rodriguez@ehu.eus (A.R.);
Asier.Fernandez@ehu.eus (A.F.); Norberto.lzlacalle@ehu.eus (L.N.L.d.L.)
* Correspondence: cfaa2015@ehu.eus or EdwarLeonardo.Sastoque@ehu.eus; Tel.: +34-688-673-836

Received: 31 October 2018; Accepted: 4 December 2018; Published: 6 December 2018

Abstract: Many manufacturing sectors require high surface finishing. After machining operations such as milling or drilling, undesirable burrs or insufficient edge finishing may be generated. For decades, many finishing processes have been on a handmade basis; this fact is accentuated when dealing with complex geometries especially for high value-added parts. In recent years, there has been a tendency towards trying to automate these kinds of processes as far as possible, with repeatability and time/money savings being the main purposes. Based on this idea, the aim of this work was to check new tools and strategies for finishing aeronautical parts, especially critical engine parts made from Inconel 718, a very ductile nickel alloy. Automating the edge finishing of chamfered holes is a complicated but very important goal. In this paper, flexible abrasive tools were used for this purpose. A complete study of different abrasive possibilities was carried out, mainly focusing on roughness analysis and the final edge results obtained.

Keywords: flexible abrasive tools; finishing; rounding edge; superalloys

1. Introduction

Titanium alloys and nickel-based superalloys are widely used today in aerospace components, commonly used in engines, considering that superalloys and concretely Inconel 718 are capable of working in corrosive environments and at high temperatures. Those materials can be used as part of gas turbine engines, steam, nuclear components, chemicals, etc. There is a strong demand for dimensional accuracy and surface roughness for these high-value components.

Drilling holes in aerospace components is often a delicate operation; the hole amplifies the stress around it by a factor of two [1]. Moreover, it is often the last machining operation, with a looming risk of making a scrap part due to a single bad hole. This circumstance determines the final time used in the production of the part, and a lack of quality can lead to its rejection, as it should especially take into account the reliability of the process due to the costs already involved. Therefore, it is a high value-added operation [2].

Currently in industrial practice, drilling processes are widely used due to their versatility and the short time invested in performing the task. However, these operations produce results of not very high quality, thus requiring complementary operations such as dotting, re-drilling, reaming, chamfering and edge finishing. This fact supposes a waste of time, both in subsequent cutting processes and in tool changes. The "not very high quality" refers, basically, to the deviations that occur in terms of diameter tolerances, surface roughness and burr formation, which are inherent phenomena in the process. Also, the effect of the cutting parameters on the hole quality (circularity and hole diameter) and tool wear during the drilling of super alloy Inconel 718 allows us infer that the cutting speed and feed rate played a large role in the variation of the deviation from the circularity values [3]. The available literature regarding drilling in high strength materials is rather limited [4,5]. However, in recent years there have been further investigations into new techniques and processes to drill holes in these alloys.

Among these new techniques, ultrasonic assisted machining is one of the most commonly used. This is a machining technology where a high frequency vibration (20 kHz) with an amplitude around 10 μm overlaps the continuous movement of the cutting tool, providing an output power between 50 W and 3000 W [6]. The use of ultrasonic-assisted processes allows a reduction in cutting forces by 30–50% [7], an improvement of the final surface quality, better chip evacuation and a longer tool life [8].

Other authors propose alternatives to traditional drilling. The idea is to use a ball-end milling tool giving it a helical motion around the hole. Regarding the helical milling, there are two similar helical milling techniques: Ball helical milling (BHM) and contouring ball helical milling (CBHM) [9]; the results were quite good in terms of quality but the times were far from those obtained with twist-drilling operations, or in other processes [10,11]. Takt-time in aeroengine manufacturing in many cases prevents the replacement of drilling with twist drills, thus edge burrs and poor finishing are common issues. In emerging processes, the plasticity of metal is also a key factor, as shown in [12,13].

In this paper, brushing techniques using abrasive flexible tools are studied. The aim is to implement these tools for the finishing process, to improve the surface finish obtained on the one hand, and to achieve the rounding of the edges in the countersunk holes on the other. Flexible hone tools are available in silicon carbide, aluminum oxide, zirconia alumina, boron carbide, tungsten carbide and even in other grades, with diameters ranging from 4 to 1000 mm.

In this work, different available state-of-the-art tools are presented. Tests were carried out in order to make a first attempt to use these tools, with interesting results that are shown below.

2. Flexible Abrasive Tools

Companies such as Brush Research Manufacturing (BRM) have a long history of solving difficult finishing problems with brushing technology. The term "brush" is commonly associated with classic twisted wire brushes or the nylon brushes used for deburring or edge blending. It is a flexible and elastic abrasive tool, ideal for soft cutting in finishing operations, "plateau honing", cylinder liner deburring, hydraulic and pneumatic components, as well as other industry sectors such as aeronautics, automotive parts, screw machining, etc.

These are a general-purpose tools (Figure 1), the versatility of which stems from the small abrasive balls overlapping at the end of a nylon filament. Each ball is independent of the others; this fact ensures the centering and auto-alignment with the hole. Having complete control of the process parameters and identifying and assessing the influence on the final surface is essential for the efficient implementation of these tools in CNC machines and robots.

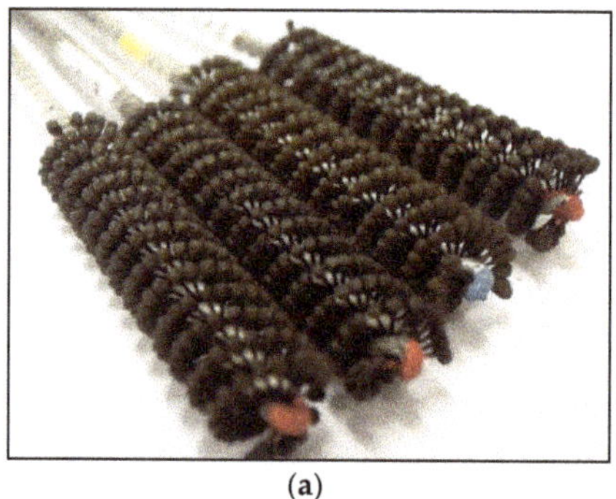

(a)   (b)

Figure 1. (**a**) Tools with different abrasive qualities. (**b**) Tool detail (3×).

One application of these flexible tools is the surface finishing and edge blending of holes made in aeronautic alloys, such as Inconel 718 and Ti6Al4V. A wide range of abrasives and grit sizes are offered by BRM and other companies. This implies the necessity to carry out a comparison between the different abrasive grades. The test results for different abrasive types and grain sizes are presented in this work. The parameter measured in this first approach was the final roughness of the brushed holes. Table 1 shows the variety of tools used. (Prices are shown because in some cases these are twice those of other solutions.)

Table 1. Different flexible abrasive tools used in tests.

Code	Abrasive	Grit Sizes	Nominal Diameter	Web Price
SC10	Silicon Carbide	180	10 mm	≈12 \$/un
SC11	Silicon Carbide	180	11 mm	≈12 \$/un
SC11-400	Silicon Carbide	400	11 mm	≈23 \$/un
BC11	Boron Carbide	180	11 mm	≈15 \$/un
Di11	Diamond	2500 C Mesh	11 mm	≈30 \$/un

3. Previous Tests on Ti6Al4V Alloy

Preliminary tests on Ti6Al4V alloy were carried out. This alloy was used firstly because is more economic, easier to buy and has better machinability than the superalloys, such as Inconel 718. Titanium plates were used ($200 \times 100 \times 7.5$ mm dimensions); the main aim of these tests was to establish a first approach to the process before carrying it out on Inconel 718. On the other hand, titanium alloys are used not only in engines but in airframe key parts, which join the wings to the airplane body.

Prior to conducting the brushing tests, 80 holes of 10.7 mm diameter were drilled in the plates. The following conditions were used for the drilling: Vc = 35 m/min and f = 0.12 mm/rev. Figure 2 shows the experimental set-up.

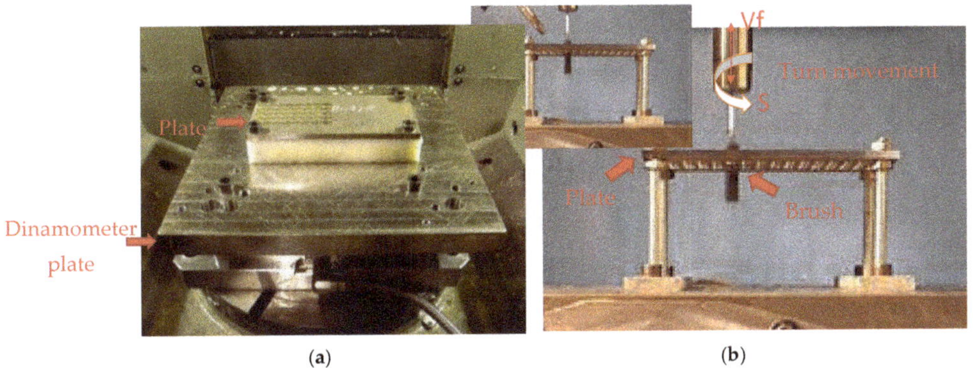

(a)   (b)

Figure 2. (**a**) Drilling set-up. (**b**) Brushing set-up.

As shown in Figure 1 and Table 1, five different flexible abrasive tools were used. With each different brush, 16 holes were made at these brushing conditions: Vc = 60 m/min and f = 0.5 mm/rev. Figure 3 shows the results of the roughness measurement, both for the drilled holes and the brushed holes. Firstly, the surface quality obtained in the previous drilling with the conditions used was quite good, averaging around 0.5 µm Ra. The main problem was the results dispersion, with varying Ra roughness parameters from 0.29 to 1.17 µm.

After brushing with the five different flexible abrasive tools, the roughness parameters decreased and the results dispersion was lower. The best brush type for this material in terms of surface roughness was the SC11-400, as it reduced the average Ra roughness up to 0.25 μm, with values between 0.2 μm and 0.3 μm (Figure 3).

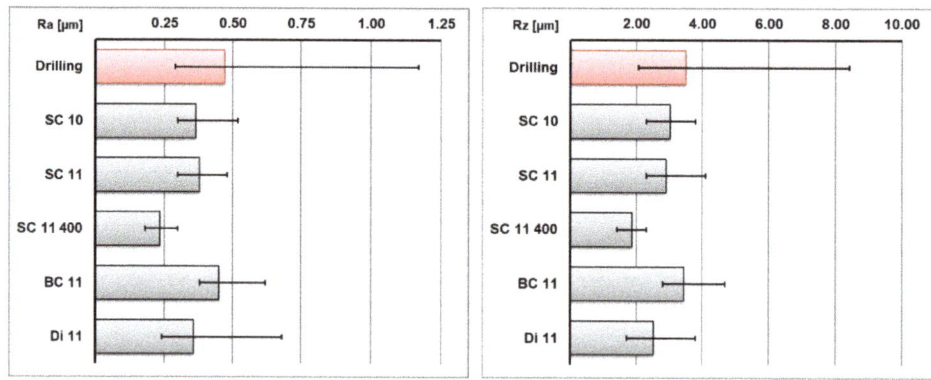

Figure 3. Ra [μm] and Rz [μm] roughness values after drilling and after finishing Ti6Al4V.

4. Test on Inconel 718 Superalloy

Based on data from the preliminary tests carried out on Ti6Al4V, we found that that the five different flexible abrasive brushes were able to reduce the roughness parameters. Moreover, it was an easy and economical finishing process that can be carried out using machine tools.

On the other hand, the preliminary tests showed a low cutting capacity. It was difficult to make a chamfer on a hole or deal with large burrs because removing that much material was impossible. However, these brushes could be useful in order to finish surfaces, round edges or carry out cross-hole deburring [9]. For these reasons, experimental tests were carried out on Inconel 718 plates, which is a commonly-used material in aerospace components working at high temperatures. This is a difficult material to machine, so the soft cut of these brushes may have been insufficient.

In this case, Inconel 718 plates were used with the dimensions $200 \times 100 \times 7.5$ mm, similar to those used in the preliminary tests in titanium. The tests were carried out in an Ibarmia ZV25 milling machine, with a spindle with 25 KWs. Regarding the previous drilling, the Table 2 shows the two different conditions used.

Table 2. Conditions used for drilling in Inconel 718.

	Vc [m/min]	f [mm/rev]	S [rpm]	Vf [mm/min]	No. Holes
"A" Conditions	20	0.06	595	35.7	40
"B" Conditions	25	0.06	744	44.6	40

4.1. "A" Conditions

Two different cutting conditions for previously-drilled Inconel 718 were tested. The first conditions were established by the tool manufacturer, the second ones were rather demanding in order to reduce the processing time and increase productivity. The aim was to compare the surface roughness results after brushing. The drilling parameters in the "A" conditions were c = 20 m/min and f = 0.06 mm/rev.

After performing the drilling, brushing tests were carried out. In this case the same brushing conditions as in the preliminary test were used (Vc = 60 m/min and f = 0.5 mm/rev).

Figure 4 shows the roughness results obtained before and after brushing. The surface roughness obtained after drilling was moderately good (around 0.5 μm Ra) thanks to the conservative drilling conditions used. After brushing, the results showed that the roughness values decreased somewhat, but not significantly. In the case of the SC10 brush, the roughness became worse. This implies that for this material and using the drilling conditions given by the manufacturer, these SC10 brushes were not suitable. As for the rest of the tests, similar to the results for the titanium, the best roughness results were achieved with SC11-400 brushes. However, the BC11 brush provided similar roughness values and less deviation in the results. In addition, the BC11 brushes were cheaper and showed less wear after brushing than SC11-400, thus BC11 was the most suitable in this case.

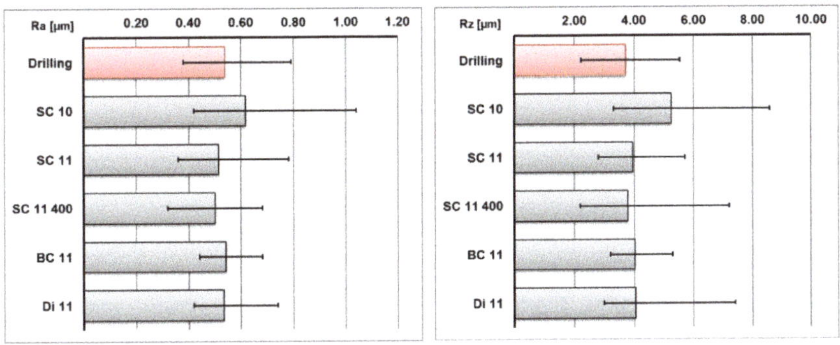

Figure 4. Ra [μm] and Rz [μm] roughness values after drilling and finishing Inconel 718.

The results obtained in this case also show that under these drilling conditions it was too unproductive to execute a brushing operation because the surface improvement was hardly noticeable. In the following section, the brushing process in more demanding drilling conditions is examined. In this case, the brushing process might be useful.

4.2. "B" Conditions

In this section, the previous drilling conditions were Vc = 25 m/min and f = 0.06 mm/rev. In this way, the roughness results obtained after drilling were worse than in the previous cases. However, brushing could be useful in this case. Figure 5 shows the roughness results. The roughness values observed before brushing were around Ra 0.9 μm, with a large dispersion of results. After brushing, the roughness parameters decreased to values lower than 0.65 μm Ra. In this case, the BC11 brush provided the lowest roughness values and the lowest deviation of the results, thus this was the most convenient brush. Besides, the tool wear was not critical in this case.

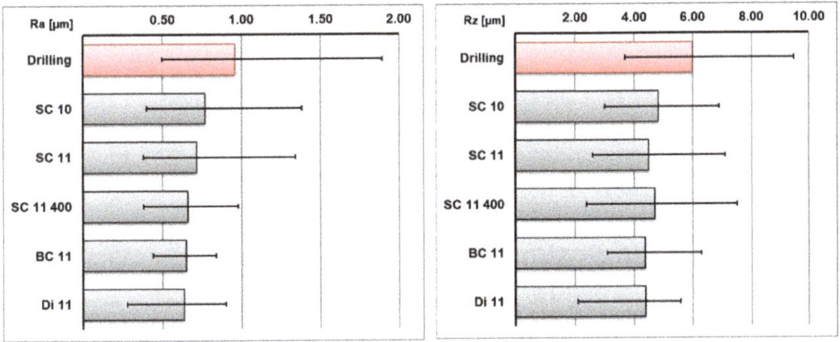

Figure 5. Ra [μm] and Rz [μm] roughness values after drilling and after finishing Inconel 718.

It is noted that the results showed that the cutting ability of these brushes was limited, especially when cutting materials with low machinability such as Inconel 718. Therefore, a large amount of deburring and chamfering of holes was impossible. However, once the hole was chamfered, rounding the edges and finishing the surface was achieved using the flexible abrasive brushes [11]. Figure 6 shows one of the drilled and brushed holes. The process involved drilling, chamfering and brushing. Figure 6 and Table 3 set out the rounding edge produced by the brushes.

(a)   (b)

Figure 6. (**a**) Hole section and detail of the chamfer and rounding edge. Image 1 (5×), Image 2 (10×), Image 3 (10×). (**b**) Angle detail measured with optical means of the rounding edge of a hole section (5×).

Table 3. Angle detail—hole section of the rounding edge.

	Angle/°	Apex X/mm	Apex Y/mm
Angle 1	102.241	99.452	48.341

For years, the edge finishing process has been by hand in many areas, but now the tendency is to try to automate these finishing processes [12]. One automation possibility uses flexible abrasive brushes. However, others are possible, such as using shape tools. In this case, the main drawback is the correct tool positioning and also the tangents to the surface. In addition, it is necessary to consider the fact that many of these holes are placed in curved areas or in areas that are difficult to access for a conventional milling tool. The main problem with the brushes is the lack of repeatability and the rapid wear suffered.

Figure 7 shows some photographs of the brushes following their use. As mentioned above, despite achieving the best results, the SC11-400 brush is one of the most expensive, along with the diamond brush. Besides, tool wear on these brushes is greater than the other brushes. To conclude, regarding tool wear, BC11 is the most appropriate option for materials such as Inconel 718. Moreover, in some cases it is also the best option in terms of the surface quality achieved.

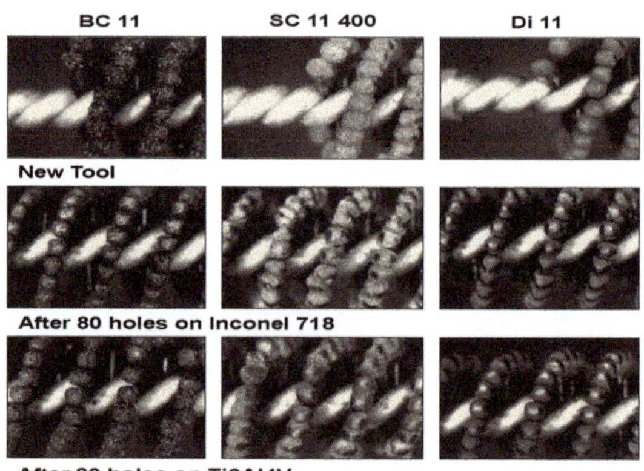

Figure 7. Tool wear on three different brushes. New tool; after Inconel 718; after Ti6Al4V (3×).

5. Automatic Process

Polishing and deburring is a process with great automation possibilities, through the use of robots [13]. Force feedback control is the key aspect to be considered. The definition of a robotic cell for the application of the process is based on accessibility. The flexible brushes being applied to holes in different aeroengine components and cases being produced with Inconel 718, Hastelloy or other nickel alloys are a good task for these robots. Deburring and edge finishing will be always a step in the process chain. The idea proposed is to work in high-automation mode, in the following stages:

- Burr detection, for instance using structured blue light or other optical means. The random location implies a random pattern.
- Robotic deburring in brush manipulation: a robotic arm can use a spindle with the usual low-torque to brush at the required rotational speed.
- Final check: optical means will help, and in cases of internal hole surface roughness, a roughness meter measurement is necessary.

The proposed system for the automation process consists of a unique superfinished cell capable of working in two different work modes, in particular with a tool on a robot or with a piece on a robot. In the first of the work modes (MOD1), the idea was to work on pieces of large dimensions (Ø 2400 mm, height 1500 mm, weight 2500 kg) mounted on a rotating table and working with tools mounted on the robot, which is able to access the outer and inner areas of the type pieces. The materials to be worked in this case will be heat-resistant alloys with mechanical characteristics equal to or higher than Inconel 718, Titanium 6-4, Jethete type stainless steel or similar. The operations to be carried out will be diverse, highlighting operations of deburring, edge killing and polishing of localized areas and holes, as well as measurement and control operations. In a second mode of work (MOD2), we will work with tools mounted in fixed posts, with the piece positioned mounted on the manipulator robot. In this case, the pieces to be treated will be units or sets coming from castings or other types of components with the maximum dimensions of approximately 1000 × 1000 × 1000 mm and maximum weights of up to 120 kg. The materials will have characteristics similar to those indicated for the MOD1 work mode and the operations to be carried out include cutting and sanding processes as well as other operations such as those already mentioned for deburring, polishing and measuring and control. Figure 8 shows the system used to apply the approach.

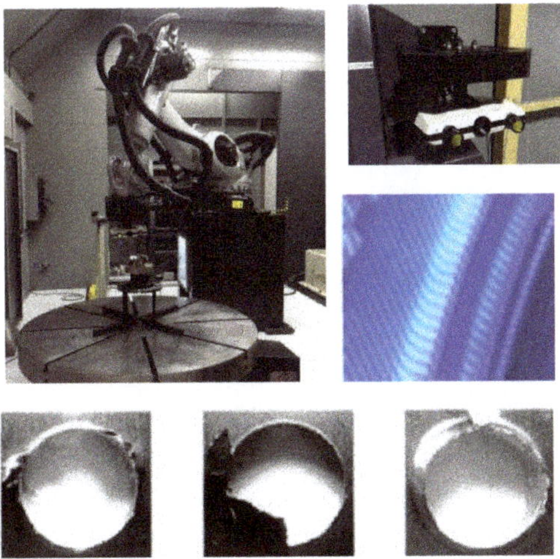

Figure 8. Robotic deburring: robot arm, structured-blue light devices, detail of holes with burrs.

6. Conclusions

Several contributions of this research can be pointed out, namely:

- Preliminary tests on Ti6Al4V show that flexible abrasive brushes are able to reduce the roughness parameters of drilled holes. Furthermore, the final roughness shows less deviation from the average value in comparison to the previously-drilled holes. In this material, considering the roughness values, Silicon Carbide 400 grit size brushes were the most suitable.

- Despite the fact that the brushes were not suitable for chamfering or removing large burrs, tests made on Inconel 718 showed that these brushes could be a great option for rounding edges and surface finishing. Particularly, BC11 brushes were the most suitable for this operation. After brushing with BC11, the roughness was better, the deviation of results was lower, and their price and wear resistance make them suitable for this aim.

- Brushes are a real choice in the industrial environment in order to achieve a rapid and efficient improvement of the inner quality of holes and to eliminate burrs at the hole edge, both at the entrance and exit of the drill bit from plates.

- Polishing, deburring, burr detection or a final check by optical means for large pieces are processes with great automation possibilities by robotic means.

Author Contributions: Conceptualization, A.R., A.F. and L.N.L.d.L.; Methodology, A.R.; Software, A.R. and A.F.; Validation, A.R. and A.F.; Formal Analysis, A.R. and L.N.L.d.L.; Investigation, A.R. and A.F.; Resources, L.N.L.d.L.; Data Curation, L.S.P.; Writing-Original Draft Preparation, A.R.; Writing-Review & Editing, L.S.P.; Visualization, L.S.P.; Supervision, A.R. and L.N.L.d.L.; Project Administration, L.N.L.d.L.; Funding Acquisition, L.N.L.d.L.

Funding: The authors gratefully acknowledge the project "Estrategias avanzadas de definición de fresado en piezas rotativas integrales, con aseguramiento de requisito de fiabilidad y productividad IBRELIABLE" (DPI2016-74845-R), and "Discos de freno premium para trenes de alta velocidad", by the Spanish Ministry of Economy.

Conflicts of Interest: The authors declare no conflict of interest.

References

1. Farid, A.A.; Sharif, S.; Namazi, H. Effect of Machining Parameters and Cutting Edge Geometry on Surface Integrity when Drilling and Hole Making in Inconel 718. *SAE Int. J. Mater. Manuf.* **2009**, *2*, 564–569. [CrossRef]

2. Tönshoff, H.K.; Spintig, W. Machining of holes: Developments in drilling technology. *Ann. CIRP* **1994**, *43*, 551–561. [CrossRef]

3. Kivak, T.; Habali, K.; ŞEKER, U. The Effect of Cutting Paramaters on The Hole Quality and Tool Wear during the Drilling of Inconel 718. *Gazi Univ. J. Sci.* **2011**, *25*, 533–540.

4. Mannan, M.; Alsagoff, S. Hole Quality in Drilling of Annealed Inconel 718. In Proceedings of the Thirty-Two NAMRC Conference, Charlotte, NC, USA, 1–4 June 2004.

5. Rui, L.; Hegde, P.; Shih, A.J. High-throughput drilling of titanium alloys. *Int. J. Mach. Tools Manuf.* **2007**, *47*, 63–74.

6. Shaw, M.C. Ultrasonic grinding. *Microtechnic* **1956**, *10*, 257–265.

7. Astashev, V.K. Effect of ultrasonic vibration of a single-point tool on the process of cutting. *J. Mach. Manuf. Reliab.* **1992**, *3*, 65–70.

8. Pujana, J.; Rivero, A.; Celaya, A.; López de Lacalle, L.N. Analysis of ultrasonic-assisted drilling of Ti6Al4V. *Int. J. Mach. Tools Manuf.* **2009**, *49*, 500–508. [CrossRef]

9. Olvera, D.; López de Lacalle, L.N.; Urbikain, G.; Lamikiz, A. New strategies for hole making in Ti-6Al-4V. *AIP Conf. Proc.* **2009**, *1181*, 361–369.

10. De Lacalle, L.L.; Lamikiz, A.; Muñoa, J.; Salgado, M.A.; Sánchez, J.A. Improving the high-speed finishing of forming tools for advanced high-strength steels (AHSS). *Int. J. Adv. Manuf. Technol.* **2006**, *29*, 49–63. [CrossRef]

11. Olvera, D.; de Lacalle, L.N.L.; Urbikain, G.; Lamikiz, A.; Rodal, P.; Zamakona, I. Hole making using ball helical milling on titanium alloys. *Mach. Sci. Technol.* **2012**, *16*, 173–188. [CrossRef]

12.   Álvarez, Á.; Calleja, A.; Ortega, N.; de Lacalle, L.N.L. Five-Axis Milling of Large Spiral Bevel Gears: Toolpath Definition, Finishing, and Shape Errors. *Metals* **2018**, *8*, 353. [CrossRef]

13.   Hameed, S.; González Rojas, H.A.; Perat Benavides, J.I.; Nápoles Alberro, A.; Sánchez Egea, A.J. Influence of the Regime of Electropulsing-Assisted Machining on the Plastic Deformation of the Layer Being Cut. *Materials* **2018**, *11*, 886. [CrossRef] [PubMed]

Journal of
*Manufacturing and
Materials Processing*

Article

Performance Comparison of Subtractive and Additive Machine Tools for Meso-Micro Machining

(Peter) H.-T. Liu [1,*] and Neil Gershenfeld [2]

[1] OMAX Corporation, 21409 72nd Avenue South, Kent, WA 98032, USA
[2] The MIT Center for Bits and Atoms, Room E15-401, 20 Ames Street, Cambridge, MA 02139, USA; gersh@cba.mit.edu
* Correspondence: peter.liu@omax.com; Tel.: +1-(253)-872-2300

Received: 1 January 2020; Accepted: 25 February 2020; Published: 3 March 2020

Abstract: Several series of experiments were conducted to compare the performance of selected sets of subtractive and additive machine tools for meso-micro machining. Under the MicroCutting Project, meso-micro machining of a reference part was conducted to compare the performance of several machine tools. A prototype flexure of the microspline of an asteroid gripper under development at NASA/JPL was selected as the reference part for the project. Several academic, research institutes, and industrial firms were among the collaborators participating in the project. Both subtractive and additive machine tools were used, including abrasive waterjets, CNC milling, lasers, 3D printing, and laser powder bed fusion. Materials included aluminum, stainless steel, and nonmetal resins. Each collaborator produced the reference part in its facility using materials most suitable for their tools. The finished parts were inspected qualitatively and quantitatively at OMAX Corporation. The performance of the participating machine tools was then compared based on the results of the inspection. Test results show that the two top performers for this test part are the CNC precision milling and micro abrasive waterjet. For machining a single flexure, the CNC precision milling had a slight edge over the micro abrasive waterjet machining in terms of part accuracy and edge quality. The advantages disappear or the trend even reverses when stack machining with taper compensation is adopted for the micro abrasive waterjet.

Keywords: meso-micro machining; micro abrasive-waterjet technology; stacking cutting; micro milling; taper compensation; flexure; subtractive machining; additive machining; micrograph

---

1. Introduction

With the development and commercialization of micro-abrasive waterjet (μAWJ) technology, supported under an NSF SBIR Phase II/IIB grant, OMAX added a MicroMAX®to its product lines of JetMachining®Centers. In collaboration with MIT Center for Bits and Atom (CBA) and Department of Mechanical Engineering, the performance of abrasive waterjet (AWJ) was compared with those of lasers, wire EDM, and CNC milling [1–4]. Reference parts including miniature butterflies, tweezers, and nonlinear load cells were selected to cut with these tools. Based on their interesting performance comparison, a MicroCutting Project was initiated with the objective to broaden the performance comparison by including several selected sets of modern additive and subtractive machine tools.

2. Technical Approach and Equipment

2.1. Technical Approach

A collection of machine tools available at OMAX, CBA, and other facilities were selected for the project. One of the early tasks was to define a reference part that was suitable to be machined and/or fabricated with all the selected machine tools. A decision was made to use one the flexures investigated at NASA/JPL as a key component of prototype microsplines for asteroid grippers developed under the Asteroid Redirect Mission [5,6]. As shown in Figure 1, the flexure consists of 11 full-length and 2 half-length spring-like elements. The length and width of the flexure elements were 36.3 mm and 0.5 mm, respectively. The separation between each element was 0.76 mm. The aspect ratio (length/width) was therefore 72.6. The widths of the element and gap between them are 0.51 mm and 0.76 mm, respectively. The aspect ratios of the length-to-width and length-to-thickness were 71.5 and 47.7. The above are the dimensions of the full-scale flexure. Small-scale flexures with 0.5, 0.4, and 0.33 were also machined using several tools. The flexure is extremely delicate and flexible. It took only very weak side force exerted onto the flexure element during machining to deflect them permanently. During the course of machining, strengthening tabs were used to support the delicate elements (Figure 1a). Figure 1b shows the tool path with several color coded curves: green—traverse without cutting, magenta—lead in and out, and blue—cut at quality five level. The tool path shown in Figure 1b included two steps: (1) machine the flexure with the tabs in place and (2) remove the tabs. Figure 1c shows the final part with the tabs removed.

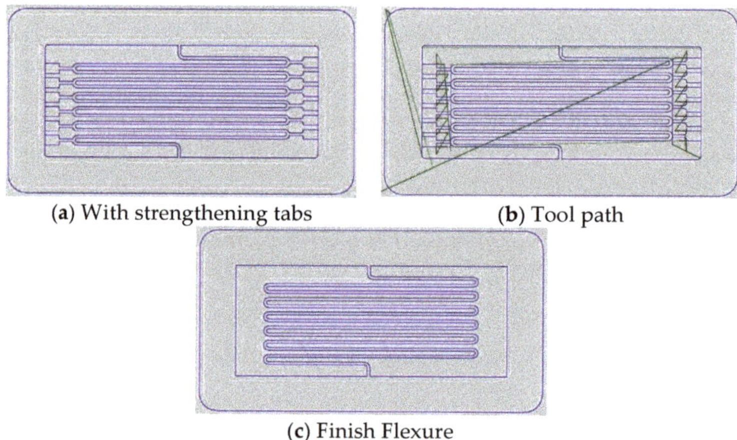

(a) With strengthening tabs   (b) Tool path

(c) Finish Flexure

Figure 1. DXF of flexure selected as the reference part for the MicroCutting Project.

The flexure, machined or fabricated from materials that were most suitable for the individual tools, were then inspected qualitatively and quantitatively to compare the performances of these machine tools.

2.2. Micro Abrasive Waterjet Technology

The AWJ is amenable to micromachining as the diameter of the AWJ can potentially reduce to micro scales [7]. The μAWJ technology was developed and commercialized under the support of an SBIR Phase II/IIB grant. Several novel processes were developed and incorporated into the MicroMAX for meso-micro machining of 2D and 3D parts. The main focus was to downsize the AWJ nozzle capable of meso-micro machining. As such, several challenges were present as the three-phase, supersonic slurry flow inside the nozzle transitions from a gravity-dominated flow to a microfluidic flow. Novel processes and apparatus were developed to enable μAWJ technology for industrial applications [4].

At present, the 7/15 nozzle with an orifice and a mixing tube ID of 0.007″ (0.18 mm) and 0.015″ (0.38 mm) is the smallest production nozzle whereas the 5/10 nozzle is currently used for special applications. Garnet with 240 mesh can be readily used with these nozzles. For garnets finer than 240 mesh, a proprietary process was developed to enhance their flowability. Success in developing the AWJ technology led to the culmination in the release of the MicroMAX as a new product debuted in 2013. It has a position accuracy and repeatability of 15 μm and 5 μm, respectively. A MicroMAX version II with the incorporation of a Rotary Axis for machining axisymmetric features was released for production in 2016.

2.3. Machine Tools and Participants

The participants in the Microcutting Project included MIT/CBA (www.cba.mit.edu), OMAX (www.omax.com), Formlabs (www.formlabs.com), Datron (www.datron.com/), (Moog Inc. (http://www.moog.com/), and BMF Precision Technology Co, Ltd. (http://bmftec.com/). Machine tools available at the facilities of the participants and used in the project included CBA Digital Fabrication Facility (http://cba.mit.edu/tools/index.html) Beam Dynamics Model LMC10000 CO_2 laser system—1.2 m × 1.2 m cutting area, 30.5 m vertical travel, 500 w (1550 W peak), 25 µm overall accuracy, 51 m/min max cutting speed (91 m/min traverse speed) Sodick SL400G Wire EDM—X, Y, Z Axis travel, 400 × 300 × 250 mm; wire diameter range: 0.051 to 0.30 mm.

Zund G-3 L-2500—Repeatability ± 0.03 mm, position accuracy ± 0.1 mm/m, working area 1800 mm × 2500 mm, high speed router 3.6 kW 50,000 rpm.

FabLight 3000 Fiber Laser—3 kW laser, working area of 6.35 m × 1.27 m and tubes of diameter 12.7 mm to 5.1 mm. Capable of cutting steel, stainless steel, spring steel, aluminum, copper, brass, titanium. Repeatability of 15 µm, accuracy of ± 20 µm/m.

Oxford Solid State Micromachining Laser—532 nm diode-pumped solid-state laser, 150 mm X-Y travel, 50 mm Z travel, 1 micron resolution. It has spot size of approximately 20 microns. The laser outputs approximately 6W of power at 10 kHz and can quickly cut through materials typically up to 0.5 mm thick. It's used for fine cutting, ablation, engraving, and marking

OMAX 5555 JetMachining Center at CBA—X-Y Travel 1.4 × 1.4 m, Tilt-A-Jet, MAXJET 5i Nozzle (10/21), 7/15 Mini MAXJET 5 Nozzle, Precision Optical Locator, pneumatic drill OMAX Corporation MicroMAX—A MicroMAX equipped with a Tilt-A-Jet (TAJ), a Rotary Axis (RA), and a Precision Optical Locator (POL) is available at the OMAX facility (https://www.omax.com/omax-waterjet/micromax). For meso-micro machining, the 7/15 and 5/10 nozzles have been used routinely for cutting demo parts and conducting in-house R&D. The nozzles were driven by an EnduroMAX 40 hp crankshaft pump (Model 4060V) with pressures up to 410 MPa. For these small nozzles, an excess flow control valve was installed to drain a part of the water through the high-pressure pump. At pressures below about 70 MPa, the Bernoulli vacuum was too weak to entrain all the abrasive into the mixing chamber. Vacuum Assist accessory was used to remove the excessive abrasive to mitigate clogging of the nozzle.

AWJ machining was controlled by an IntelliMAX Software Suite featuring an extensive tool set to streamline production. The Suite is based on a precision cutting model in which each engineering material is assigned with a machineability index according to its properties derived from the results of extensive cutting tests. The intuitive Suite that is easy to use consists of a PC-based CAD-LAYOUT and CAM-MAKE. A set of steel slats spaced 25 mm apart is installed inside each JetMachining Center to support the workpiece. A 10 cm thick polyurethane honeycomb placed on top of the slats is often used to provide a firm support to workpieces made from thin materials. Since the AWJ exerts very low force onto the workpiece, it can be secured by relatively simple fixtures such as carpenter clamps. A number of options is available to secure the workpiece depending on the setup. For small parts, thin tabs are incorporated into the tool path to prevent losing it into the tank below. Machining is carried out by setting a standoff distance of 0.76 mm between the tip of the nozzle to the top surface of the workpiece. The tool offset is set to on half of the exit diameter of the AWJ. Depending on the thickness of the workpiece and the required edge quality of cut from Q1 for raw cut to Q5 for precision cut, the cutting speed is set intelligently by MAKE according to the machineability index. The cutting speed also varies according to the shape and curvature of the tool path. For example, the traverse of the nozzle speeds up in straight segments and slows down during corner passing to maintain the same kerf width.

Formlabs-Form 2 Printer—Build Volume: 145 × 145 × 175 mm, Layer Thickness: 25, 50, 100 microns, Laser Spot Size: 140 µm, Laser Power: 250 mW, Wavelength: 405 nm (violet), automated resin system, self-heating resin tank, auto-generated supports. Machine size: 350 mm (L) Ā—330 mm (W) Ā—520 mm (H).

Moog, Inc. —These parts were made from metal powders, an aluminum alloy and stainless steel, using laser powder bed fusion. The equipment included:

- EOS 290 for 17-4PH stainless steel: laser power—220 watts, volume scan speed—750 mm/s, volume hatch spacing—0.11 mm, layer thickness—40 μm;
- SLM 280 for aluminum (twin laser): laser power—350 watts, volume scan speed—1650 mm/s, volume hatch spacing—0.13 mm, layer thickness—30 μm.

They were then trimmed to correct thickness with a ϕ0.25-mm wire EDM (Mistsubishi MV2400S) for 30 min approximately.

Boston Micro Fabrication (BMF) Material Technology Inc.—Digital Lighting Processing (DLP), similar to Stereolithography Appearance (SLA), was used to fabricate the flexures (https://web.archive. org/web/20140221025534/, https://thre3d.com/how-it-works/light-photopolymerization/digital-light-processing-dlp). It is a 3D printing process working with photopolymers. In DLP, a 3D model is constructed and 'sliced' through software. Once the sliced images are received by the printer, curable liquid, e.g. monomer or pre-cured resin, is exposed to a pattern of UV-light, in order to selectively solidify a cross-section of the designed parts. The cured cross-section is then lowered below the surface level of the liquid, allowing the liquid to backfill for curing and bonding of subsequent cross-sections. The process is repeated until all the slices of the 3D model and hence the printing parts are completed. The liquid is then drained from the vat, followed by demolding and post-curing of the parts. The nanoArch Micro Scale 3D Printing System InP140 (https://bmftec.com/) nanoArch®is the first commercialized high resolution, multimaterial 3D micro-fabrication equipment based on PμLSE (Projection Micro Litho Stereo Exposure) technology, which is designed for scientific R&D of functional composite materials.

Datron-Neo Milling Machine (https://www.datron-neo.com/us/datron-neo-simple-milling/overview/). Masking tapes and super glue were used to secure the workpiece onto the substrate.

- For the full-scale flexure, the first operation was a 2D contour used to cut the profile of the internal flexure geometry using a double flute ϕ0.030" (0.76 mm) end mill with ethanol coolant. Ethyl alcohol, used in minimal quantities, appeared to have no negative effect on the work holding. The last operation was cutting the perimeter with a 3 mm single flute end mill.
- For the half-scale flexure, the first operation was a 2D contour used to cut the profile of the internal flexure geometry using a double flute 0.5 mm end mill with ethanol coolant. Ethyl alcohol, used in minimal quantities, appeared to have no negative effect on the work holding. Datron tooling was used, which led to slightly undersized internal geometry. The internal geometry was 0.008" (0.20 mm) instead of the 0.010" (0.25 mm) as modeled due to the larger tool size being used. The last toolpath was cutting the part out with the 3 mm single flute end mill.

3. Results

For this project, one of the flexures as a key component of a prototype microspline of an asteroid gripping device under development at NASA/JPL for the Asteroid Redirect Mission [5,6]. The delicate geometry of the selected flexure, as shown in Figure 1, presented considerable challenge to most machine tools. There is an option to (1) machine or fabricate the flexure with the strengthening tabs attached (Figure 1a) and then remove the tabs to finish the part or (2) machine the flexure without the tabs (Figure 1b).

3.1. Waterjet Cutting

The prototype flexure made of 6061 T6 aluminum using the MicroMAX was originally cut to demonstrate its performance versus that of the wire EDM. The EDM process conducted at JPL was carried out in three passes in order to minimize the damage resulted from the induced heat-affected zone (HAZ). The comparison showed that the cost ratio for machining the part was 14:1 in favor of the waterjet, leading to a 93% cost reduction [3].

Figure 2a shows the micrograph of the aluminum flexure (0.81 mm THK) cut on the MicroMAX using the 7/15 nozzle with the Barton 240 or 220UT mesh garnet with a mean particle size of 60 μm. The pump pressure was 380 MPa and the abrasive mass flow rate was 73 g/min. The cutting time was 2.3 min. The geometry of the flexure element including the horizontal and semi-circle segments were inspected under the microscope. From the micrographs, the following features of the flexure are inspected:

a.   Edge quality

-       Smoothness and the presence of chipping
-       Edge taper

b.   Variation in the element width (over or under cutting)
c.   Variation in the gap between horizontal segments
d.   Straightness (local bending) of horizontal segments
e.   Parallelism of the horizontal segments

Figure 2b shows the superimposition of the tool path onto the flexure element. The overall match between the tool path and the flexure element is excellent, indicating that there is no distortion of the flexure element in terms of bending and/or rotating in the X-Y plane.

Magnified views of the two small areas in the middle span and the right end loops of Figure 2a, were selected to compare in detail the flexure element and the tool path, as shown in Figure 2c,d. The areas were chosen because they were farthest away from the two anchoring points and least supported. The selection was made to show the worst mismatches in those two segments of the entire element. Yet, the match shown in Figure 2c is excellent. A very slight mismatch is observed in Figure 2d. The maximum mismatch was measured to be about 0.1 mm, part of which is attributed to the error in overlaying of the tool path onto the micrograph. Figure 2c,d show nearly no macro distortion induced by the waterjet cutting process, clearly demonstrating the advantages of cold cutting and low-side-force exertion. In fact, previous investigation verified that waterjet, as opposed to wire EDM, preserved the structural and chemical integrity of parent materials [8].

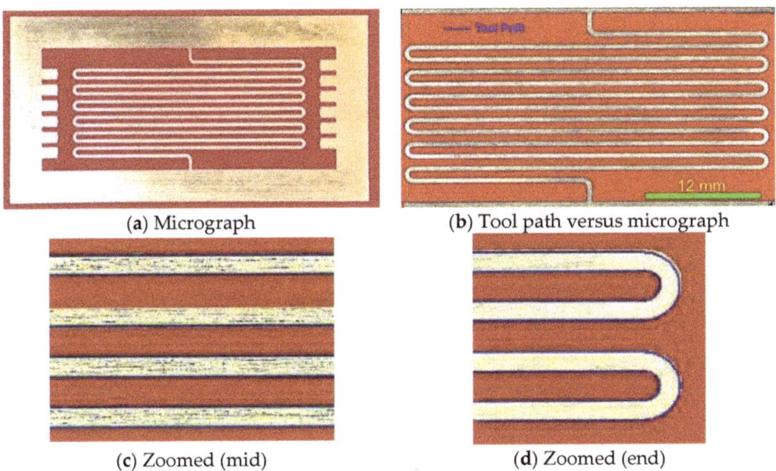

(**a**) Micrograph   (**b**) Tool path versus micrograph

(**c**) Zoomed (mid)   (**d**) Zoomed (end)

Figure 2. Aluminum flexure—MicroMAX.

The same flexure was then cut from a stainless-steel sheet (0.76 mm THK). The same cutting parameters for cutting the aluminum flexure were used. The cutting time increased to 3.4 min. Figure 3 illustrates the micrograph of the stainless-steel flexure. The flexure element displays no heat- and/or mechanically-induced distortion. Using the 7/15 nozzle, a 2/3-size but not a half-size flexure was successfully cut as the kerf width of the 7/15 nozzle is larger than the gap between elements of the half-size flexure.

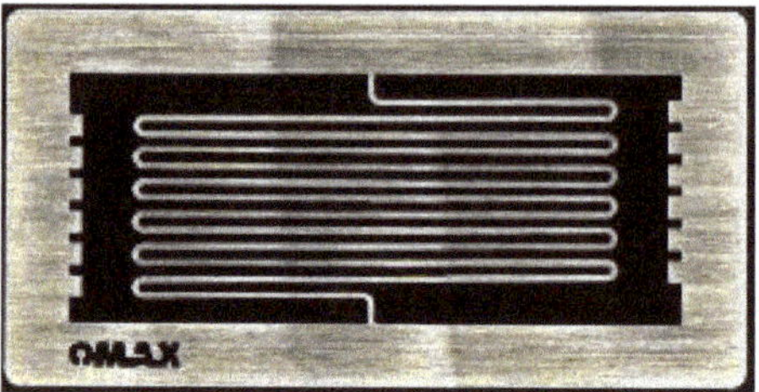

Figure 3. Stainless steel flexure—MicroMAX.

Full-size flexures were cut on CBA's OMAX 5555. The same setup for the 7/15 nozzle was used. Figure 4 shows micrographs of two flexures cut on aluminum (0.81 mm THK) and stainless steel (0.64 mm THK). In Figure 4a,b the overall geometry shows no noticeable distortion of the flexure element. Minor mismatches between the tool path and the flexure element are however observed in the zoomed-in micrographs of the aluminum flexure as shown in Figure 4c,d that correspond to the mid-span and right end segments of the second and third loops from the top. In the mid-span (Figure 4c), an undercut section about 9 mm long was observed on one of the horizontal segments (third row down. The maximum undercut is about 0.15 mm. Since this is the only occurrence throughout the entire flexure element, it is most likely just an outliner. The maximum mismatch at the two ends is about 0.1 mm. As shown in Figure 4d, the Y-positions of the first loop are slightly below its designed positions marked by the tool path. In other words, this corresponds to the small rotational (clockwise) distortion about the axis of the X-Y plane. Note that no such rotational distortion was observed on the part cut with the MicroMAX (Figure 2d). Comparison of Figure 4c,d with Figure 2c,d indicated that the cutting accuracy is slightly better for the MicroMAX than for the OMAX 5555. This is expected as MicroMAX was specifically designed and constructed for meso-micro machining.

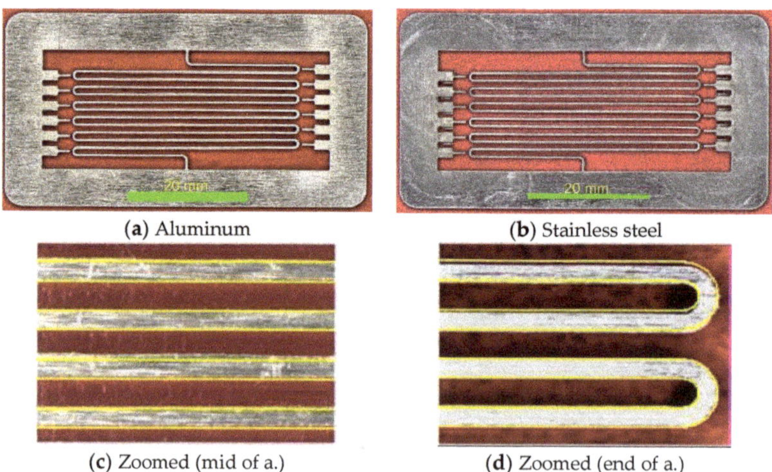

(a) Aluminum	**(b)** Stainless steel
(c) Zoomed (mid of a.)	**(d)** Zoomed (end of a.)

Figure 4. Metal flexures—5555 JMC (CBA).

3.2. Zund

Aluminum flexure (2024 0.51 mm THK) was machined on the Zund G-3 L-2500. Machining was carried out using a 0.76 mm diameter end mill with amorphous diamond coating (Harvey Tool 72030-C4). The end mill was driven by a 50 krpm spindle. A faced aluminum sheet was used to provide a level and rigid support to the stock. PSA tapes burnished with a small stainless-steel rod were applied to both surfaces of the support and the underside of the stock. The tapes were then bonded with CA glue.

The cutting time was 2.5 minutes that is comparable to the waterjet on the aluminum sheet 1.6 times thicker. Figure 5 shows the micrographs of the flexure (a), superposition of tool path flexure element (b), and the corresponding zoomed sections (c and d). Figure 5b shows that the overall geometry of the flexure matches well with the tool path. Magnified views of two segments of the mid-span and left end of the fourth and fifth loops from the top are shown in Figure 5c,d (flipped horizontally). They reveal that there are minor mismatches between the flexure element and the tool path. The maximum mismatch is about 0.1 mm. Figure 5d shows that the mismatches were also attributed to rotational distortion.

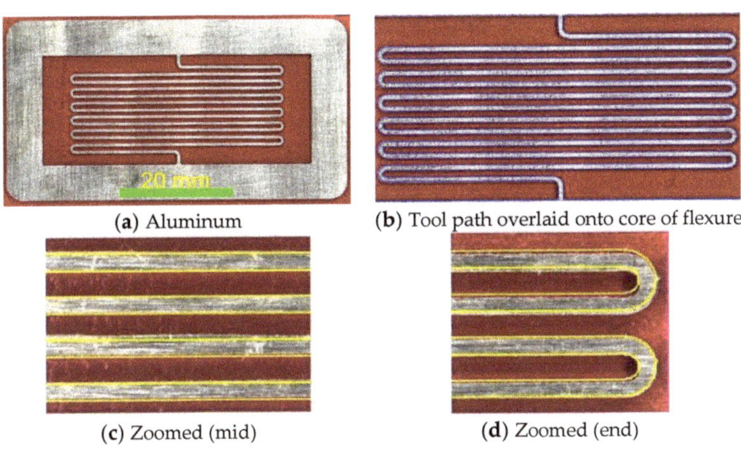

(a) Aluminum   **(b)** Tool path overlaid onto core of flexure

(c) Zoomed (mid)   **(d)** Zoomed (end)

Figure 5. Aluminum flexure—Zund (CBA).

By comparing the magnified micrographs of flexures machined with the two waterjets and Zund, the match between the tool path and the flexure element appeared to be slightly better for the one cut with the MicroMAX than the one cut with the Zund whereas the match between the tool path and one cut with the OMAX 5555 ranked third. For the flexure machined on the MicroMAX, the mismatch was mostly attributed to undercutting at the end loops. All the horizontal segments of the flexure element show no observable rotational distortion about the axis perpendicular to the X-Y plane. For the flexure machined on the Zund, the mismatches were attributed to the rotational distortion of the several segments of the element about the axis perpendicular to the X-Y plane. Note that there were differences in the fixturing setup. During waterjetting, tabs were used to strengthen the flexure element. They were then cut away after machining of the flexure element were completed. On the other hand, the Zund machined the flexure without tabs. The support was provided by the PSA tapes bonded together with the CA glue.

A 0.5 scale flexure (2024 aluminum 0.51 mm THK) was machined on the Zund by using a 0.38 mm diameter end mill with amorphous diamond coating. It took about 8 minutes to cut that part. Figure 6a shows the half-size flexure. There was only a negligible distortion of the flexure element. One of the most important attribution to the success in cutting the delicate flexure was the use of the diamond coated end mill. According to Toress et al. [9], the fine grain nanocrystalline diamond (NCD) coating, with average grain size 30–300 nm, reduced the thrust and main cutting forces to less than 50%, compared with uncoated tools, when machining 6061 T6 aluminum. One of the failure modes was delamination of the coating. After delamination, the tool looks similar to an uncoated tool with worn and/or broken tool corners and adhered workpiece material. End mills coated with the NCD experienced delamination about one half of those coated with the fine-gained diamond (FGD) counterpart. At present, the materials most suitable for the Zund are limited to relatively soft materials (At present, aluminum is the only metal recommended by Zund (http://fab.cba.mit.edu/content/tools/zund/manual.pdf).

0.5 scale aluminum flexures were also machined on the MicroMAX. A beta 5/10 nozzle with IDs of a diamond orifice and a mixing tube, 0.13 and 0.25 mm, was used to machine the flexures. Several cutting parameters were set to pump pressure, 345 MPa, garnet 320 mesh at a flow rate 45 mg/min. For the powdery 320 mesh garnet, however, it did not flow well under gravity feed. A proprietary process and a novel feeding apparatus were developed to enhance flowability of the fine powdery garnet under gravity feed (patented). Figure 6b shows the waterjet-cut half-size flexure on 6061 T6 aluminum. Its element shows no observable distortion.

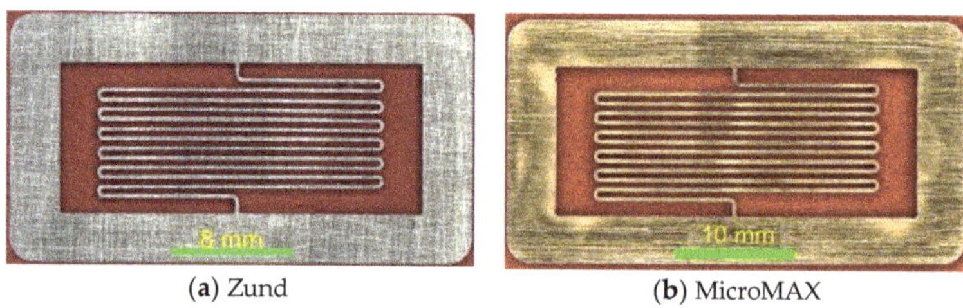

(**a**) Zund   (**b**) MicroMAX

Figure 6. 0.5 scale aluminum flexures.

Subsequently, an experimental 4/8 nozzle with a ϕ0.1 mm diamond orifice and a ϕ0.20 mm mixing tube was used to machine 0.4 scale flexures. Most of the cutting parameters were kept the same except that the mass flow rate of the 320 garnet was changed to 34 mg/min. 0.58 mm thick 304 stainless steel was used as the material. Figure 7 illustrates the micrographs of 0.5 and 0.4 scale flexures machined with the 5/10 and 4/8 nozzles, respectively. Both showed little visually observable distortion.

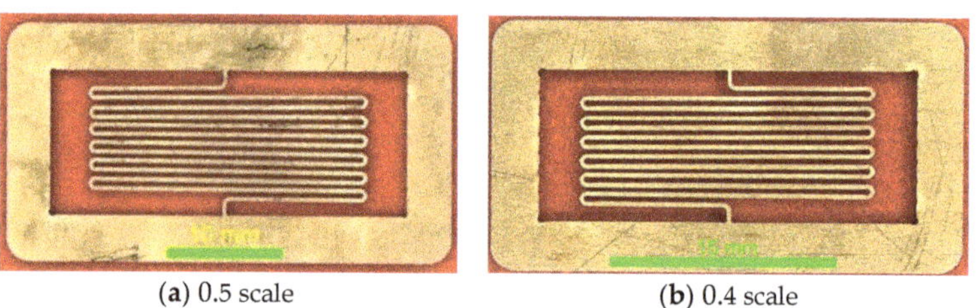

(**a**) 0.5 scale   (**b**) 0.4 scale

Figure 7. µAWJ-cut 0.5 and 0.4 scales stainless steel flexures (0.58 mm thick).

Attempts to use the 4/8 nozzle to machine a 0.33 scale failed. At that scale, the gap between the flexure element was 0.25 mm while the minimum achievable kerf width cut with the 4/8 nozzle is 0.23 mm. The cross-section of flexure element was 0.17 mm (width) $\times$ 0.58 mm (thickness). In principle, the 4/8 nozzle should be able to machine the 0.33 scale flexure. It turned out that the large-aspect-ratio flexure element was insufficiently stiff to maintain its shape.

3.3. Laser Machining (FabLight Fiber Laser)

The FabLight laser was used to cut a flexure on a 304 stainless steel sheet (0.61 mm THK). The laser was operated in the 2-pulsed mode. The cut and pierce parameters are listed in Table 1. Figure 8a,b show the back side of the as-cut and cleaned flexures. The cutting speed was too slow so that heat melted materials on the cutting edges of the workpiece. The molten metal flowed downward and re-solidified on the lower surface forming the recast or slag. The side view of the flexure would look just like that on a CO_2 laser-cut parts [4]. One of the remedies was to increase the pulsation frequency from 500 Hz to 5 kHz at the sacrifice of the cutting speed.

Table 1. Cut and pierce parameters of FabLight laser.

Cut Parameters	Speed mm/s	Power %	Gas Pressure MPa	Height mm	Frequency Hz	Focal Offset mm
	2.5	15	0.41	0.76	500	−0.76
Pierce Parameters	Pulses	Power %	Pressure MPa	Height mm	Frequency Hz	Duration ms
	1	10	0.41	1.52	500	12

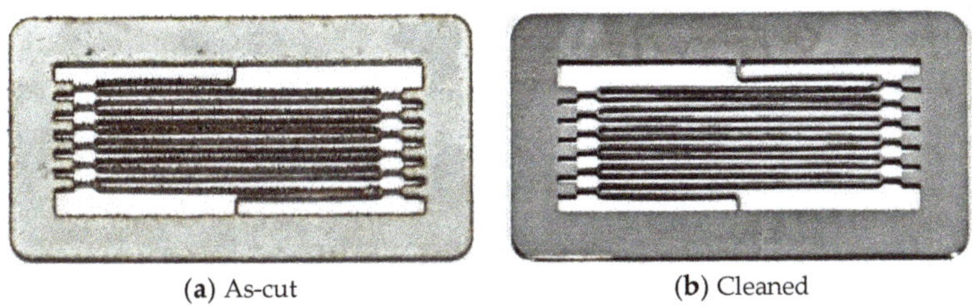

(**a**) As-cut   (**b**) Cleaned

Figure 8. Stainless steel flexure—FabLight Fiber Laser (CBA).

3.4. Laser Powder Bed Fusion-LPBF (Moog)

At Moog, flexures made from metal powders, AlSi10MG and 17-4 stainless steel, were fabricated using laser powder bed fusion (LPBF). Table 2 lists the process parameters. The flexures were then cut to desire thickness using wire EDM. Flexures with and without the tabs were built. Figure 9 shows the two 0.51 mm thick aluminum flexures (9a with tabs and 9b without tab) and a 0.64 mm thick stainless steel (9c. without tab). The surface pattern observed in the figure was left behind by the wire EDM trimming. From the high-resolution micrographs, the width of the elements is consistent. There is however minor distortion of the elements in terms of bending are observed, resulting in small variations in the gap width between the elements. The degree of distortion is less for the ones with tabs. Without the tabs, the distortion is less for the 17-4 build than for the aluminum counterpart. The density of metals produced via LPBF is typically 99.7% that of (fully dense) wrought material [10]. The reduction in density is due to various forms of material defect, which are typically small i.e. less than 0.10 mm. These defects in concert with a rough as built surface finish can result in reduced fatigue performance when compared with smooth surfaced wrought material. For example, the S-N (stress versus number of cycles to failure) curves for LPBF-built AlSi10MG using the ALB1 process show that the average fatigue life of specimens printed at 0, 30, 60, and 90 degrees is about 65% of that of the aluminum wrought [11]. The fatigue life of the SLM (selective laser melting)-built steel 630 is about 57% that of its wrought [9]. It has been demonstrated that the fatigue life of SLM AlSi10Mg parts can been extended to about 8% by machining and heat treatment [12]. Tests will be needed to determine whether 3D printing would be suitable to fabricate flexure for the intended application. It should be pointed out that the fatigue lives of AWJ-cut aircraft aluminum and titanium were able to extend considerably through dry-grit blasting on the part edges [13,14].

Table 2. Metal flexures—laser powder bed fusion.

Metal Flexures	Aluminum Alloy	Stainless Steel
Powder Materials Used	AlSi10Mg	17-4 PH Stainless Steel
Equipment-Manufacturers	EOS 290	SLM 280, twin laser
Laser Power (W)	350	220
Volume Scan Speed (mm/s)	1650	750
Volume Hatch Spacing (mm)	0.13	0.11
Layer Thickness (μm)	30	40
Part Thickness (mm) §	0.51	0.64
Wire EDM-Mistsubishi MV2400S	Wire Dia. 0.25 mm, ~30 min per part	
Build Time (hrs)	~5	~5
Heat Treatment ¥ (°C/hrs)	300/3	1150/1.5
Tools for Removal of Support Structure	Pliers, custom fixture and pneumatic cutoff wheel	

§ Used wire EDM to trim to desired part thickness. ¥ For stress release only.

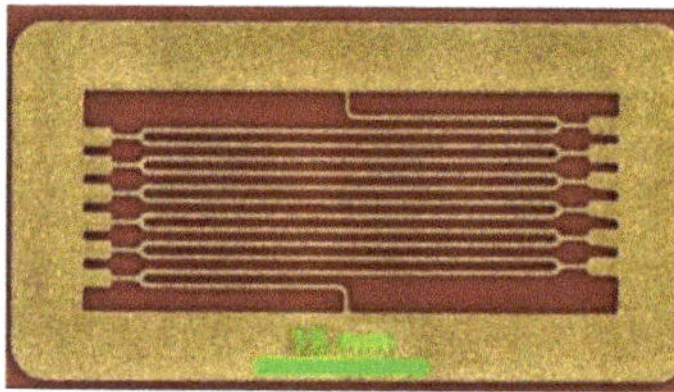

(**a**) AlSi10Ng with tab (0.51 mm THK)

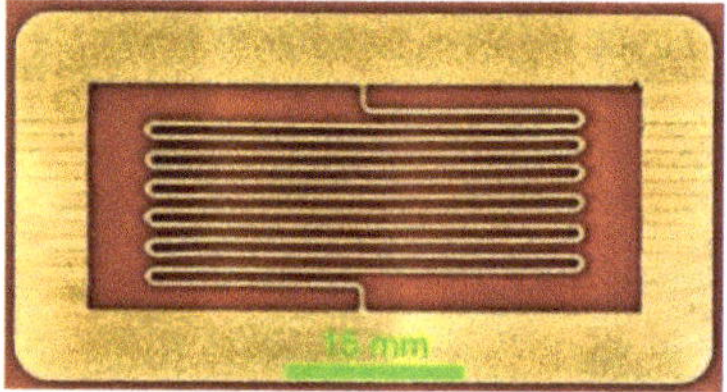

(**b**) AlSi10Ng without tab (0.51 mm THK)

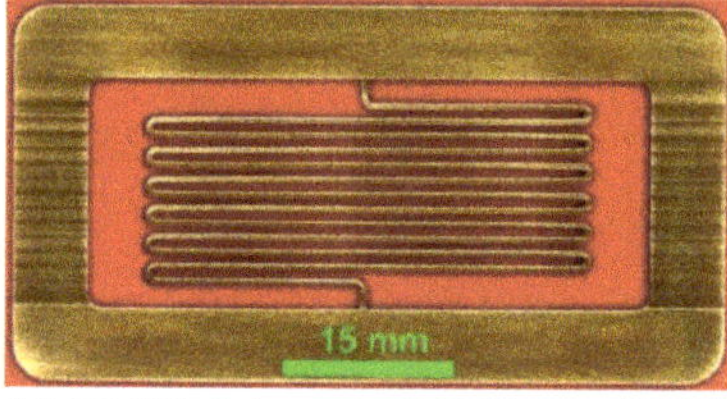

(**c**) 17-4 Stainless steel (0.64 mm THK)

Figure 9. Metal flexures—LBPF.

3.5. 3D Printing (NanoArch Micro Scale—BMF InP140/InS140)

Flexures were printed at BMF Material Technology, Inc. in Shenzhen City, Guangdong Province, China. Flexures were printed using two materials, GR and HTL, that were acrylic based photosensitive resin developed by BMF. Refer to its material properties in Table 3 (https://bmf3d.com/materials/). Two GR flexures (0.51 mm and 1.02 mm THK) are shown in Figure 10a,b. The surfaces of the flexures were quite smooth and flat. Both showed noticeable distortion along the direction of the *Y*-axis on the X-Y plane. A portion of the element segments were bent slightly as indicated by the nonuniform gap width between several straight sections of the element. The distortion is more severe on the thick flexure than on the thin one. Figure 10c shows a third flexure built from the HTL material in black color. Distortion of the flexure element was also observed.

Table 3. Material properties of GR and HTL.

Resin	GR (Hard)	HTL
Tensile Strength	85 MPa	79.3 MPa
Elasticity Modulus	3.8 GPa	4.2 GPa
Elongation at Break	3%	2.23%
Bending Strength	97.4 MPa	120.6 MPa
Flexure Modulus	3.2 GPa	3.96 GPa
Impact Strength	47.5 J/m	30 J/m
Distortion Temperature 45 MPa	102 °C	140.7 °C

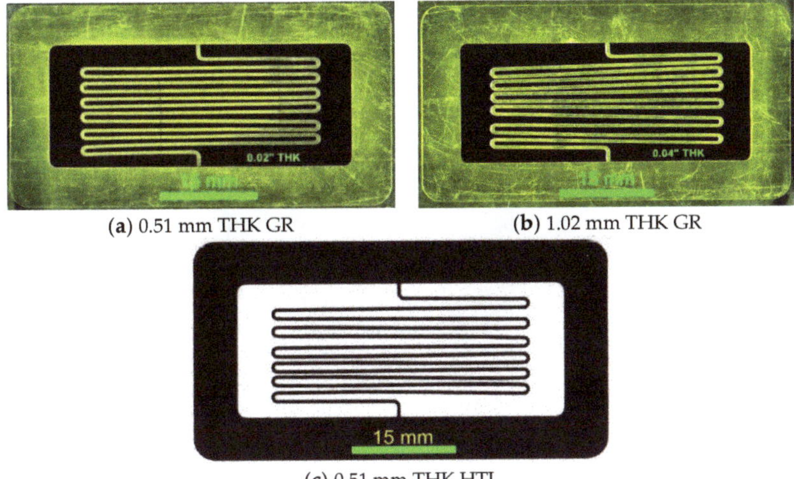

(**a**) 0.51 mm THK GR

(**b**) 1.02 mm THK GR

(**c**) 0.51 mm THK HTL

Figure 10. 3D-printed flexures (BMF).

3.6. 3D Printing (Formlabs—Form 2)

Stereolithography, an additive manufacturing process that polymerizes a liquid resin with light, was used at Formlabs to print two different sets of flexure with different supporting structures [15]. A Model Form 2 printer, a galvanometer system to steer a laser on a cure plane for this purpose, was used in this case. A model is sliced into layers as thin as 0.025 μm and created layer-by-layer on this cure plane (https://formlabs.com/blog/ultimate-guide-to-stereolithography-sla-3d-printing/). Wherever the laser hits the resin, the material hardens into the final part. An inverse stereolithography process, parts are formed "upside down", and are drawn up from a tank full of rigid resin that was reinforced with glass to offer very high stiffness and polished finish [16].

Figure 11 illustrates two flexures built with the Form 2. The horizontal and radial elements of both flexures display considerable distortion. According to Formlabs, the peel and squish forces of the print process were most likely responsible for the distortions. After each layer, the part separates ("peels") from the tank. This motion is a combination of the tank moving laterally and the Z-axis moving upwards. After separation, the part then returns to its original position, though one-layer-thickness higher. For small fragile parts with long thin features, this separation and return can generate forces that cause the part to return slightly off of position. Since there are lots of thin features next to each other, over time this displacement added up enough to cause them to get close enough such that the liquid resin around them caused them to stick together through viscous forces such as surface tension, enough to hold them in place.

3.7. Micromachining (Datron)

High-speed CNC milling was used to machine two flexures (full and 1/2 scales) at Datron. They were cut on a double flute end mill using ethanol coolant on a NEO CNC Machine equipped with the Autodesk Fusion 360 software. The material was 2024 aluminum 0.51 mm thick. The aluminum sheet was secured with masking tape and super glue. The cutting parameters were given in Table 4.

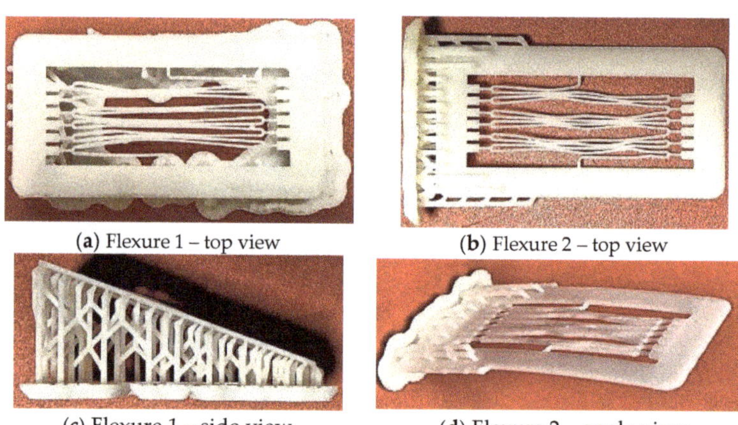

(**a**) Flexure 1 – top view	(**b**) Flexure 2 – top view
(**c**) Flexure 1 – side view	(**d**) Flexure 2 – angle view

Figure 11. Flexures built with rigid resin—3D printing (Formlabs).

Table 4. Cutting parameters for micromachining with Datron Neo CNC Machine.

Scale (%)	Article No.	Ø End Mill (mm)	RPM (×1000)	Feed XY (mm/min)	Feed Z (mm/min)	D.O.C. [¥] (mm)	W.O.C. [§] (mm)	Cycle Time
100	N/A	0.76	38	1016	254	0.127	0.76	6 min 1 s
50	0068005KK	0.48	38	305	127	0.076	0.48	16 min 12 s

[¥] Depth of cut; [§] Width of cu.

Figure 12 shows the two flexures. There is no apparent distortion on the horizontal and radial segments of the full-scale flexure element. However, there is observable distortion on the top three horizontal segments with non-uniform gap spacing. It was noted that workpiece holding issues prevented optimization of speeds and feeds. A very shallow depth of cut and slow feed rate was required to prevent tool breakage. An uneven application of the superglue underneath the masking tape could cause tools to break. This allowed for slight movement of the workpiece during machining. It was also very difficult and time consuming to remove the tape and super glue on the flexure without damaging the delicate part.

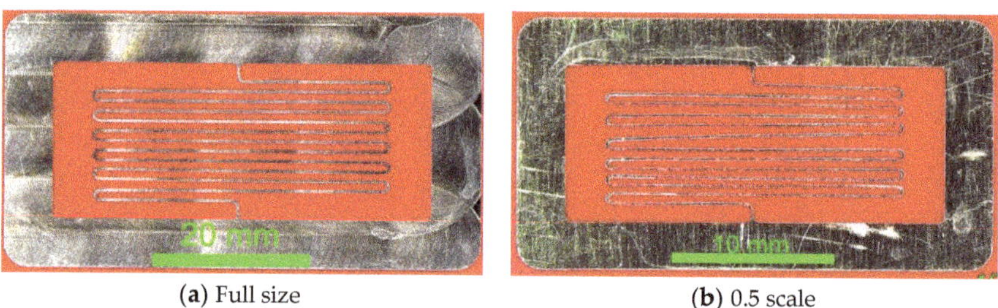

(**a**) Full size	(**b**) 0.5 scale

Figure 12. Aluminum flexures—Neo at Datron.

Micrographs of the half-scale flexures machined with the Zund, MicroMAX, and Neo (at Datron) are replotted side-by-side in Figure 13. Figure 13a1,b1,c1,a2,b2,c2,a3,b3,c3 show the as-cut full flexures, cores of the flexure elements overlaid with the tool paths, and the magnified views of the mid-span and two end loops of the flexures, respectively. Visual comparison of the micrograph of the waterjet-cut flexure and the tool path shows no degradation (column b) resulted from the downsizing. The maximum deviation is still about 0.1 mm. Since there is a slight edge rounding on the jet entry surface of the flexure, the micrograph shown in column b corresponds to the jet-exit surface of the flexure. In the presence of the edge taper, the width of the flexure element is slightly but consistently wider than the tool path in the presence of the edge taper. For the Zund-cut counterpart, however, the downsizing has led to certain degradation in the match between the flexure and the tool path. Figure 13a3 displays noticeable rotational distortion. The maximum mismatch was measured to be 0.27 mm, nearly three times that for the full-scale flexure. Considerable rotational distortion is observed for the Neo-cut half-scaled flexure, as shown in Figure 13c3. The maximum mismatched was measured to be 0.66 mm.

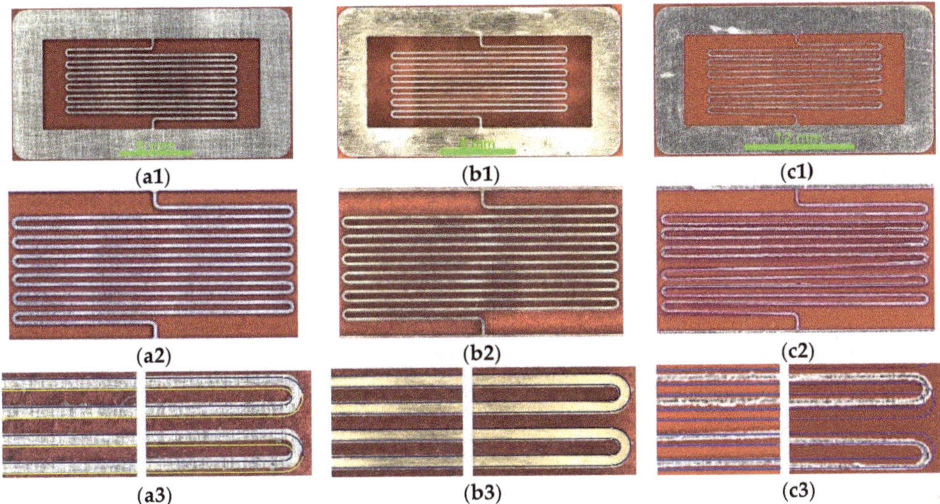

Figure 13. 0.5 scale fluxures cut with Zund, MicroMAX, and Neo at Datron (from left to right): As-cut flexures with frames (**a1**, **b1**, and **c1**); tool paths superimposed onto flexure elements (**a2**, **b2**, and **c2**); and tool paths superimposed onto flexure elements—zoomed in (**a3**, **b3**, and **c3**).

3.8. Micgraphs of Flexure Elements

The 0.5-scale flexures cut with the Zund, MicroMAX, and Datron (Neo) were further inspected under a microscope to compare the performance of the three machines for micromachining. Figure 14 illustrates typical micrographs of top and bottom views of a single end loop cut with the three tools. Note that the designed width of the flexure element and the gap between the horizontal segments of the element are 0.25 and 0.38 mm, respectively. The sum of the two is 0.634 mm. From the micrographs, we measured the average values of these dimensions. The corresponding values for the flexures machined with the Zund, MicroMAX, and Datron are (0.240, 0.387 mm), (0.262, 0.366 mm), and (0.156, 0.462 mm), respectively. Comparison of the dimensions of the flexures machined with the three tools is summarized in Table 5. Note that the gap is governed by the diameter of the cutting tool while the sum of the measured element width and gap is the same as that of the designed dimensions.

Table 5. Comparison of flexure dimensions *.

Reference and Tools	Jet/Tool Diameter (mm)	Width of Flexure Element (mm)	Gap Between Element (mm)	Sum of Element and Gap Width (mm)	Comment
Designed	N/A	0.25	0.38	0.63	
Zund	0.38	0.24	0.39	0.62	
MicroMAX (4/8)	0.25	0.26	0.37	0.63	
Datron (neo)	0.48	0.16	0.46	0.62	

* Average value derived from micrographs based on 10 to 20 measurements.

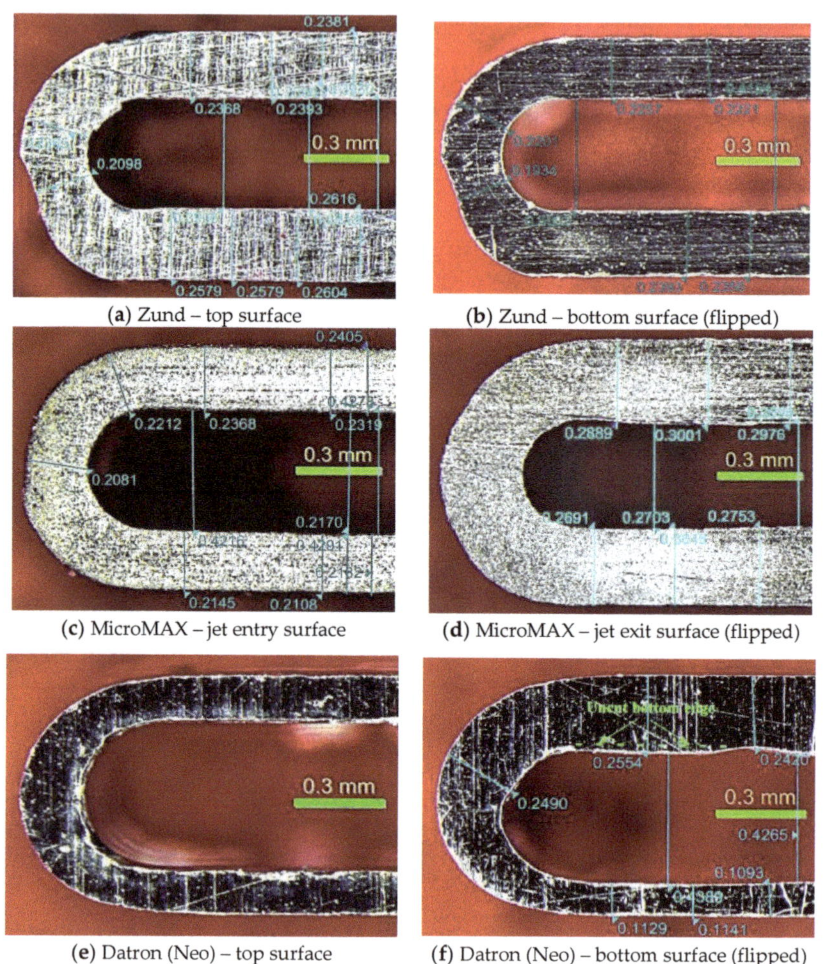

Figure 14. Visual comparison of micrographs of top and bottom surfaces of half-scale flexures cut with Zund, MicroMAX, and Neo at Datron.

In Figure 14 each part displays certain anomalies. For example, there is a consistent flat spot on the outside loop of the Zund-machined flexure. For the MicroMAX-machined flexure, a minute rounding on the edge of the jet entry surface and an edge taper can be observed. The average edge taper was measured to be 33 µm. The corresponding edge taper for the Zund- and Neo-machined flexures are considerable smaller, that is 8 µm. Note that waterjet and the end mill of CNC milling are a flexible and a hard tool, respectively. During machining, waterjet bends and spreads and its cutting power reduces as it cuts into the workpiece. As a result, a natural taper forms on the waterjet–cut edges. On the other hand, the end milling is in direct contact with the cut edge of the workpiece. The minute edge taper is likely caused by the deflection of the miniature end mill. The JetMachining Center is equipped with a 5-axis accessory, Tilt-A-Jet (TAJ), capable of compensating edge taper (https://www.omax.com/accessories/tilt-a-jet). The TAJ is however not effective in machining thin materials; it will be applied to stack cutting in Section 3.9.

Comparison of these values with the designed dimensions indicates that the flexures machined with the Zund and the MicroMAX match well with the designed dimensions. The large deviations of the element width and the gap on the flexure cut with the neo at Datron is attributed to the large diameter of the end mill (0.48 mm). Figure 14f shows that the Neo did not cut through the materials at several spots (below the green dashed line). The poor performance of the Neo is partly attributed to the imperfect fixturing to secure the workpiece, according to Datron. As a result, the workpiece might have been moved during machining. Using better tape and more even application of super glue, similar to the process used on the setup of the Zund, would likely allow an increase to depth of cut and feed rate, reducing the cycle and handling times of this operation.

3.9. Measured Width of Flexure Elements

The widths of the flexure elements were measured from the micrographs to compare the cutting accuracy of the μAWJ on the MicroMAX and the CNC milling on the Zund. Figure 15 shows the micrographs of the top and bottom views of flexure elements machined on the MicroMAX and Zund, respectively. Comparing Figure 15a,b shows that the bottom elements are slightly wider than their top counterparts. This is the result of the presence of edge taper as the TAJ was deactivated for cutting thin materials. There are two options to reduce or minimize the magnitude of the edge taper. One remedy is to reduce the cutting speed and the other is to conduct stack cutting. The OMAX MAKE software incorporates an optimal stack height calculator.

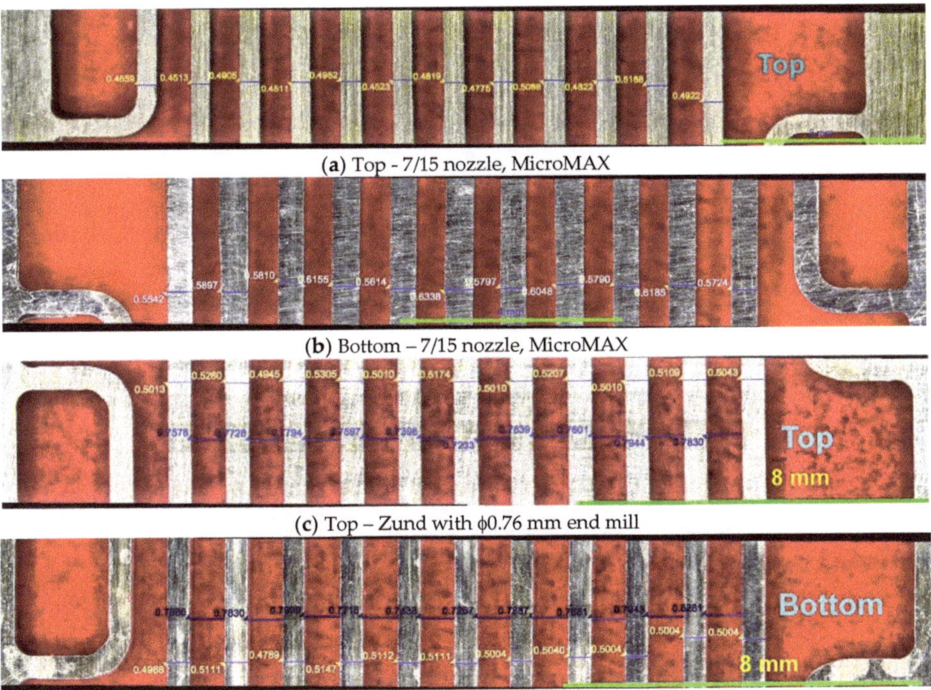

(a) Top - 7/15 nozzle, MicroMAX

(b) Bottom – 7/15 nozzle, MicroMAX

(c) Top – Zund with ϕ0.76 mm end mill

(d) Bottom – Zund with ϕ0.76 mm end mill

Figure 15. Width of flexure elements measured from micrographs.

The average widths of the top and bottom flexure elements were measured from the micrographs and are presented in Figure 16. Also shown in the figure are their trendlines and the designed width of the flexure elements. The maximum deviations between the trendlines and the designed width are 0.004 mm and −0.0003 mm for the μAWJ- and Zund-cut flexures. The measured width of the flexure machined with both tools displays a consistent pattern with the even segments wider than the odd counterparts. The difference in the width between the odd and even elements is considerably larger for the μAWJ-cut flexure than for the Zund-cut one. Careful examination of the tool path shown in Figure 1b indicates that, for waterjet machining, the nature of the machining differs for the odd and even segments of the flexure elements. Specifically, the odd and even elements were cut in the constrained and unconstrained modes, respectively.

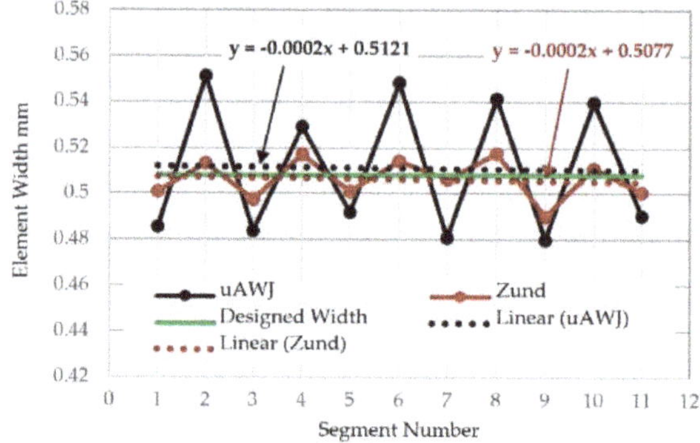

Figure 16. Average element widths of full-scale flexure (C-C and U-U) machining mode).

Unconstrained or constrained cuttings referred to the conditions that the edges of the adjacent elements had already or yet to be cut. Note that the tool offset for the 7/15 nozzle is one half of the mixing tube diameter, or 0.38 mm whereas the gap between elements are 0.76 mm. While cutting the odd segments, the AWJ is constrained by the materials on both edges (C-C). The AWJ cut nearly straight downward. On the other hand, for the even segments with the edges of the adjacent segments already cut, the AWJ is unconstrained on both edges (U-U). This is referred to as the constrained-constrained and unconstrained-unconstrained (C-C/U-U) cutting mode. Under the U-U cutting mode, the AWJ tends to be deflected slightly toward the adjacent segments where the material beyond those edges have been removed. As such, the even segments are slightly wider than their odd counterparts. Refer to Figure 17 for a graphical interpretation of the two modes of cutting.

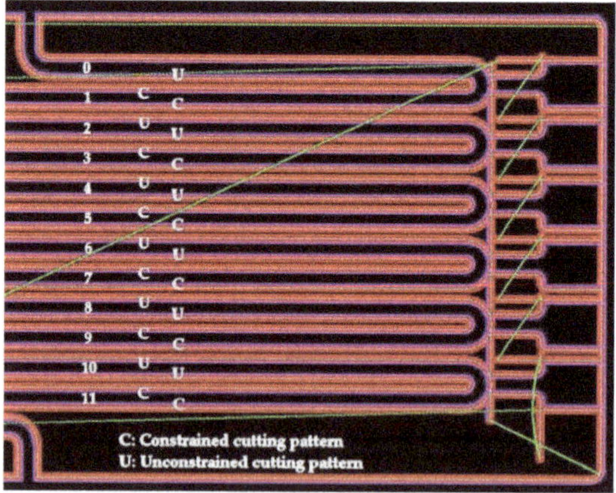

Figure 17. Tool path for the C-C and U-U cutting mode.

The remedy to mitigate the difference in the element width is to revise the tool path such that the machining must be carried out consistently in the unconstrained-unconstrained (U-U) or alternating constrained-unconstrained (C-U) modes for all the elements. For the unconstrained mode, a slotting in the middle of the gap between elements, as illustrated in Figure 18, is added to the tool path (magenta dotted lines). As such, all the edges are carried out in the unconstrained mode. The second remedy is to split the tool path of the flexure element into two sub-segments at the left turn-around points as shown in Figure 19. Cutting are conducted by cutting all the top edges in the constrained mode followed by cutting the bottom edges in the unconstrained mode. As such, the edges will be cut in the C-U mode.

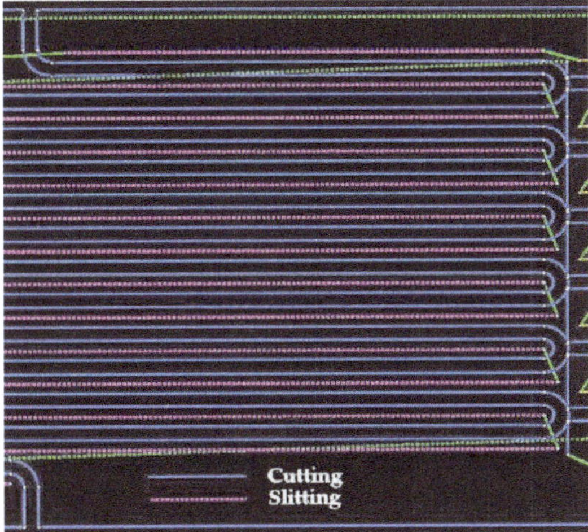

Figure 18. Tool path for U-U cutting mode.

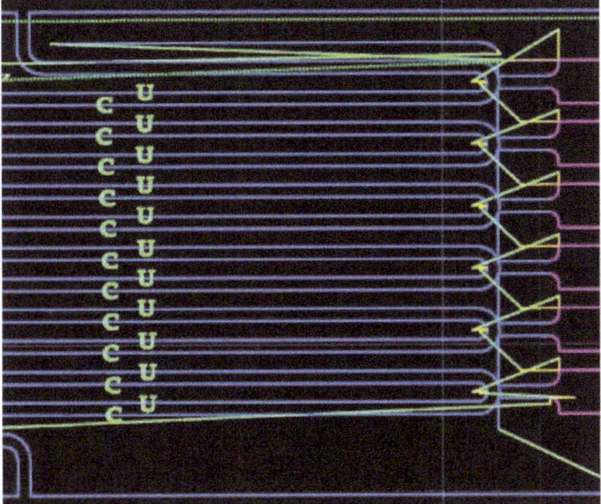

Figure 19. Tooth path for C-U cutting mode.

Cutting tests using both modes were conducted on 0.51-mm thick aluminum. The difference in the width of the odd and even flexure elements reduced significantly and were about the same as that cut with the Zund, as shown in Figure 20. However, the pre-slitting along the middle of the gaps significantly weakened the stiffness of the workpiece, particularly toward the mid-span. During the passage of the AWJ, the force exerted onto the workpiece tended to push the flexure elements sideway and downward, particularly at mid-span where the support was the minimum. As a result, the element width became nonuniform along the length of the flexure element. In particular, the width increased from one end, reached the maximum at the mid-span, and then reduced toward the opposite end, as shown in Figure 21. The same trend with less severity was observed for the C-U cutting mode.

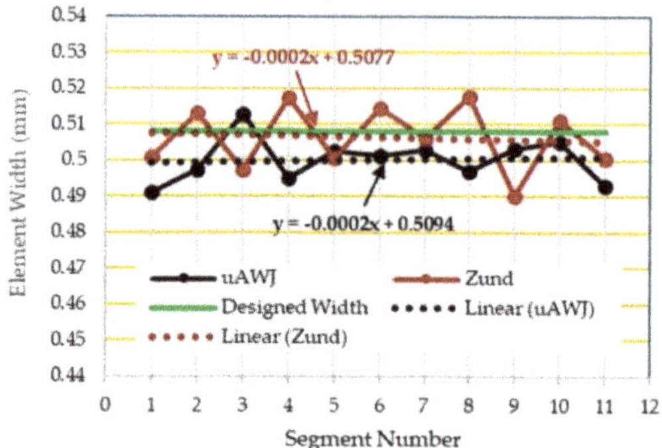

Figure 20. Comparison of average element widths of flexures cut with µAWJ (U-U mode) and Zund.

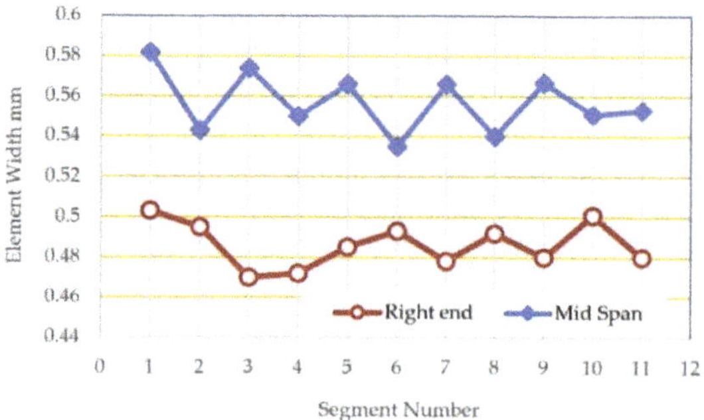

Figure 21. Nonuniform element width resulted from the C-U cutting mode.

The above anomalies of the observed nonuniformity of the element width along its major axis is attributed to the weak stiffness of the flexure element due to its peculiar geometry.

A set of flexure elements with large aspect ratios of length/width and length-/thickness.

The stiffness of the serpentine flexure supported only on two end points connected to its frame is the weakest at the mid-span.

For cutting very thin materials, the cutting speed is too fast for the TAJ to respond with taper compensation. Therefore, the TAJ is usually deactivated resulting in measurable edge taper. Figure 22 shows the measured element width at the mid-span of a 0.51 mm thick stainless-steel flexure under the C-C/U-U cutting mode. In addition to the nonuniform element width of the even and odd flexure segments, the presence of the edge taper resulted in a difference of about 0.1 mm between the element widths measured on the top and bottom surfaces.

Subsequently, we decided on adopting the AWJ stack cutting process as the final remedy to minimize the nonuniformity of the width of the even and odd flexure elements and the edge taper. The stack was formed by using 3M double-adhesive tapes with a thickness of about 60 µm. A 10 mm thick aramid honeycomb with fiberglass faceplates was used to support the metal stack for further enhancing the stiffness. The cutting model is equipped with a stack calculator to estimate the optimum number of layers based on the cutting parameters. The optimum number of sheets is defined as the total stack height at which the average cutting time per sheet has reached the minimum. Figure 23 illustrates the results of stack calculation for two aluminum (0.51 and 0.64 mm thick) and one stainless steel (0.51 mm thick), respectively. The estimated numbers of layers for the three metals cut at the quality level of five are 13, 10, and 7, respectively. The cutting times for single sheets reduce from 3.90

to 0.90, 4.13 to 1.12, and 4.95 to 1.80 min, respectively. The reductions in the cutting time are 4.4, 3.7, and 2.8 times for the three metals.

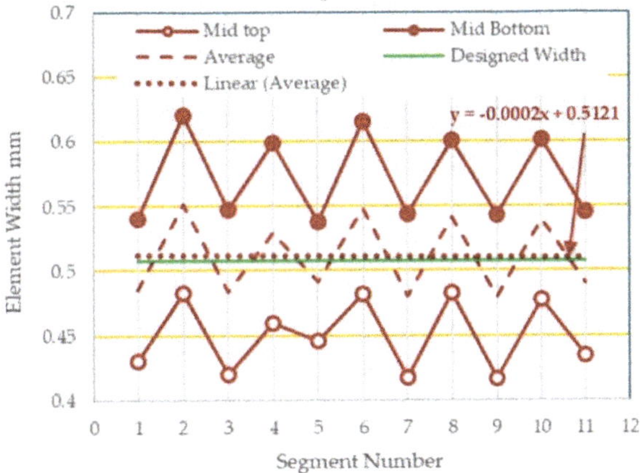

Figure 22. Edge taper resulted from deactivation of the TAJ.

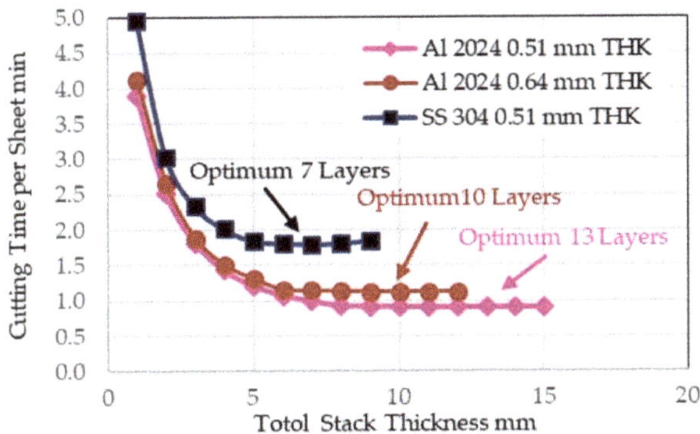

Figure 23. Typical optimum stack heights for two metals.

Figure 24 show the side view of a stack of flexure consisting of one 0.41 mm thick aluminum sheet and five 0.51 mm thick stainless-steel sheet. The top aluminum sheet served as the sacrificial cover where the edge rounding, and frosting took place. Several iterations of cutting with the TAJ activated to minimize the edge tape were conducted. Figure 25 illustrates the element widths of one of the interior flexures (#5) of the stack shown in Figure 24. The significant reduction in the edge taper of the μAWJ-cut flexure was evident when compared the data shown in Figures 24 and 25. Although the edge taper was still slightly higher than that of the Zund-cut counterpart it can further reduce with additional iterations. Most important, the stack cutting eliminated the difference in the width of the even and odd elements, even below the level achievable with the Zund.

Figure 24. µAWJ-cut stack of stainless-steel flexure supported by aramid honeycomb.

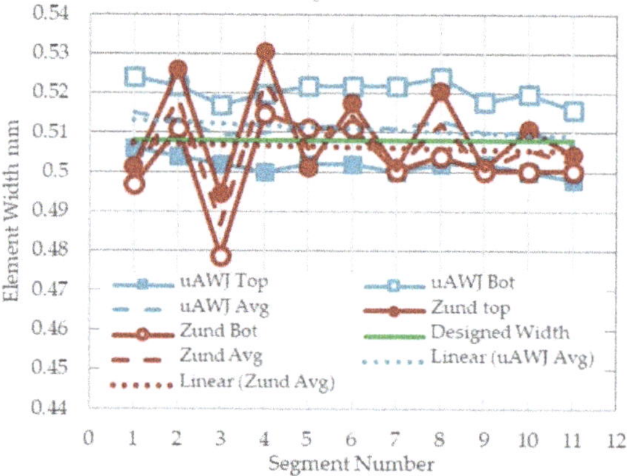

Figure 25. Comparison of element width of flexures cut with Zund and µAWJ under stack cutting at mid-span.

With the increase in the stiffness of the stack together with the support of the honeycomb, the sideway and downward displacements of the flexure element in response to the µAWJ loading also reduced considerably. As a result, the nonuniformity of element width reduced accordingly, as shown in Figure 26. In fact, comparing Figures 26 and 27 shows that the total variation in the element width was less for the µAWJ-cut flexure than the Zund-cut counterpart.

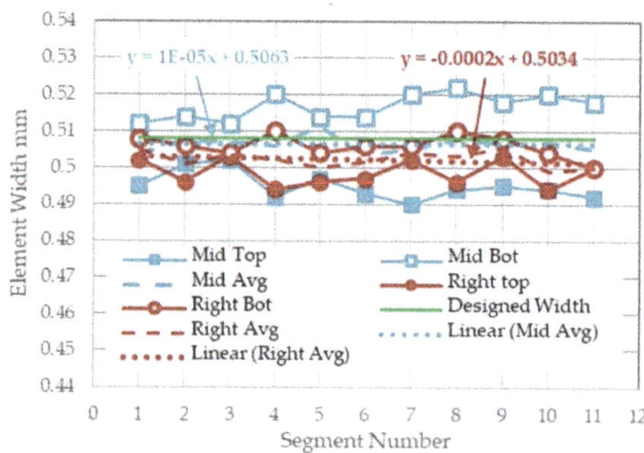

Figure 26. Element width measured at mid-span and one end of flexure cut with µAWJ under stack cutting.

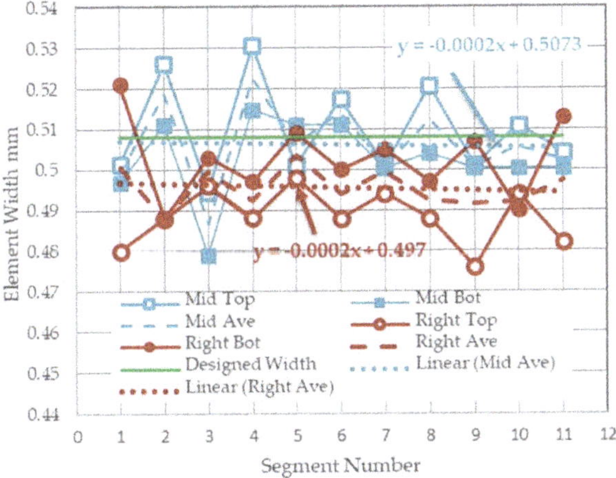

Figure 27. Element width measured at mid-span and right end of single-sheet flexure cut with Zund.

Yet another advantage of stack cutting is to use the top sheet as the sacrificial cover to reduce frosting on the top surface and the burr on the bottom edge of the interior sheets.

3.10. Further Downsizing of µAWJ Nozzle

As discussed in Section 3.2, machining a 0.33 scale flexure was unsuccessful using the 4/8 µAWJ nozzle. One of the reasons was that the kerf width of the nozzle was nearly the same as the width of the gap of the flexure. During machining the µAWJ traversed twice (back and forth) through the gap. In the presence of edge rounding on entry side, the strength or stiffness of high-aspect-ratio flexure element might be weakened to the degree that it could no longer maintain its shape without distortion. Further downsizing the µAWJ nozzle might be needed to machine the 0.33 scale flexure successfully.

Attempts were made to assemble a 2/6 µAWJ nozzle to machine the flexure. The length of the ϕ0.15 mm mixing tube was reduced to 12.7 mm. Specially processed 320 mesh garnet with a flow rate of 30 mg/min was used. The vacuum assist option was activated to boost the low Venturi vacuum induced by the small waterjet. Figure 28 shows the comparison of three stainless steel flexures, with scale of 0.5 (a1–a3), 0.4 (b1–b3), and 0.33 (c1–c2), machined on the MicroMAX with the 5/10, 4/8, and 2/6 nozzle, respectively. As discussed in Section 3.2 (Figures 2 and 5) and Section 3.7 (Figure 13a,b), the matches between the tool paths and the full-and 0.5-scale parts were slightly better for the µAWJ–cut flexures than for the Zund-cut counterparts. Figure 28c3 shows that the overall match between the 0.33-scale flexure and the tool path displayed a slight localized degradation when compared with the matches with its larger counterparts.

Considerable R&D is being conducted to continue downsizing µAWJ nozzles toward micromachining. The material independent waterjet is capable of machining a wide range of part size and thickness [1,2]. One of the main concerns is the lack of proper fixturing devices to hold extremely thin stocks for micromachining. The success in applying AWJ stack cutting to stiffen the workpiece while enhancing the cutting efficiency has eased the above concern and paved the way for precision AWJ micromachining of very thin stocks provided further development of µAWJ technology would meet the stringent requirements. A wide range of materials from metal, nonmetal, to anything in between can be used to form the stack [2].

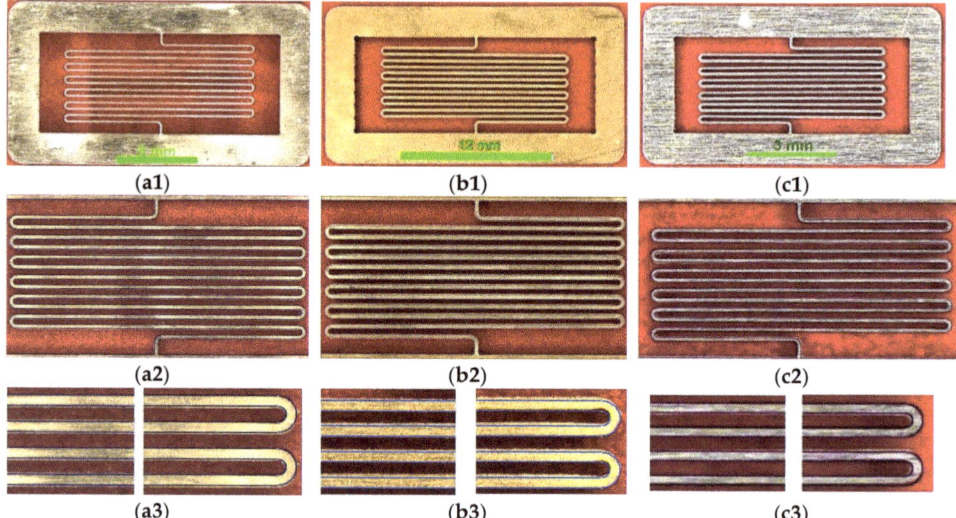

Figure 28. Downsized stainless steel flexures cut with µAWJ nozzlles on MicroMAX: flexure with frames (**a1**, **b1**, and **c1**); tool paths superimposed onto flexure elements (**a2**, **b2**, and **c2**); and tool paths superimposed onto flexure elements—zoomed in (**a3**, **b3**, and **c3**).

4. Discussion and Summary

Cutting tests were conducted to investigate the performance comparison of µAWJ, lasers, wire EDM and CNC milling. These tests were investigated through the collaboration of MIT and OMAX Corporation [4]. The results demonstrated that µAWJ using the MicroMAX had the best overall performance for this test part, with the fastest cutting speed without inducing heat damage to the parts. The CO_2 laser performed the worst causing significant heat damage (i.e., the presence of the HAZ) in terms of discoloring, warping, and the presence of excess slag. The solid-state laser pulsed at 50 kHz with a spot size of around 50 µm and the wire EDM with a 0.15 mm wire were able to cut the parts at significantly slower speed to minimize the heat damage. The cutting speeds were one to two orders of magnitude slower than that of the MicroMAX. The cutting accuracy of the solid-state laser and the wire EDM are however higher than that of the µAWJ that had jet diameters of 0.25 and 0.3 mm for the 5/10 and 7/15 nozzles, respectively.

Under the MicroCutting Project, one of the flexures used as the prototype microsplines for the NASA asteroid gripper was selected as the reference part for all the machine tools investigated. It must be clarified that such a selection may not necessarily take advantage of the best features of some of the machine tools. Specifically, the rating of the performance of individual tools was based narrowly on the inspection of the as machined/built flexures. In other words, the machine tool with a poor rating in this report does not represent its overall performance for other machining applications.

Micrographs or photographs of most finished flexures machined with individual tools were taken for inspection. The graphs were inspected and compared to determine the performance of individual tools for machining and building of the flexures. The performance was rated based on several criteria such as the cutting accuracy and speed, degree of part deformation (mechanical and thermally induced), edge quality, setup time and effort, and others. One of the inspection methods used frequently was to superimpose the part tool path onto the graph of the flexures as the means to determine whether there was any mismatch of the two. A mismatch could be caused by 3D part distortion induced by the cutting tools, inaccurate machining and building, and other factors. Another inspection method was to measure various dimensions of the flexure such as the width and the length of the flexure element and the spacing between elements.

3D printing using nonmetals, such as the GR and HTL from BMF and polymer with solid filler from Formlabs, although precisely fabricated, do not have the strength and stiffness to maintain the shape of the flexure without distortion in terms of warping, bending, and deflection. Furthermore, the printing processes usually took hours to complete. AlSi10Mg and 17-4PH stainless steel flexures built with LPBF using metal powders and finished the flexures with wire EDM by Moog Inc. appear to maintain their shape well. They also took several hours to build. Due to the presence of defects and voids in LPBF-built materials as compared with the wrought and a relatively rough surface of the finish parts, their fatigue performance is likely to be negatively impacted. The performance of the LPBF-built parts is expected to improve progressively as the process continues to refine.

Thermally based manufacturing processes such as lasers and wire EDM can potentially induce heat damage resulting from the induction of the HAZ. The remedy is to reduce the cutting power by pulsing the lasers at high rates or cutting the part with EDM at multiple passes, at the expense of the cutting time [4].

The test results show that the width of the flexure element is sensitive the mode of cutting for waterjet. The original tool path consisted "constrained" and "unconstrained" cutting modes for the odd and even flexure segments, respectively. As a result, the width of the odd segments is slightly but consistently narrower than that of their even counterpart. One of the remedies was successfully implemented by modifying the tool path such that the two edges of each element were cut under the constrained (C) and unconstrained (U) modes, respectively. The large-aspect ratios of the flexure element in both length-to-width and length-to-thickness had very low stiffness. The flexure elements were displaced sideway and downward in response to the forced exerted by the μAWJ. The stiffness was the weakest at the mid-span and increases toward the two ends. Test results showed that the element was wider at the mid-span than at the two ends, indicating that the material removal was inversely proportional to the amplitude of the displacement.

Subsequently, AWJ stack cutting was applied to improve the performance of the μAWJ. A stack of several pieces of aluminum and/or stainless steel was assembled by using a 3M double-sided adhesive tape. The total thickness of the stack was about 4 mm that was thinner than the optimum thickness estimated by the optimum stack height calculator resided in MAKE. Test results demonstrated that AWJ stack cutting has achieved the following improvements:

1. Stacking increases the stiffness of the workpiece and minimizes the lateral and downward displacement in response to the force exerted by the μAWJ. The variation in the flexure width along the its axis has reduced to the level comparable or better than that of the Zund under single sheet cutting;

2. Stacking increases the overall thickness of the workpiece enabling the activation of the TAJ for taper compensation. As such, the edge taper of the flexure reduces significantly and is comparable to that achievable with the Zund under single sheet cutting;

3. There is considerable potential for further downsizing μAWJ technology toward micromachining of most materials including nanomaterials [2,4,5,17–20]. One of the challenges is the proper fixturing of extremely thin and delicate materials to facilitate micromachining. Stack cutting together with honeycomb support has paved the groundwork for μAWJ machining of such materials.

In conclusion, several sets of subtractive and additive machine tools were applied to fabricate a reference part, a prototype flexure developed at NASA/JPL as components of microsplines on asteroid grippers for the Asteroid Redirection Mission. Aluminum and stainless-steel flexures with scales from full, 0.5, 0.4, to 0.33, were fabricated. Only the AWJ using experimental micro nozzles was able to fabricate flexures with 0.5 scale and smaller. The performances of the selected tools were evaluated qualitatively and quantitatively, as summarized in Table A1 in the Appendix A. It should be pointed out that these tools may not be optimized for fabricating the reference part. Based on the test results, the performances of the μAWJ on the MicroMAX platform and the CNC micro milling conducted on the Zund G-3 L2500 stood out among all the tools investigated in the MicroCutting Project. For machining a single piece of flexure, the Zund performed slightly better than the MicroMAX in terms of part accuracy (element width and the uniformity along its axis) and edge quality (roughness and taper). When stacking together with taper compensation using the TAJ was adopted for the μAWJ, the above advantages disappeared or the trend even reversed. The combined stack machining and taper compensation not only improved the part accuracy and edge quality but also enhanced the productivity of the μAWJ. Comparing to single-sheet machining, optimum stacking reduced the cutting times to about 4 and 3 folds for aluminum and stainless-steel sheets, respectively. As mAWJ is further downsized toward micromachining of very thin and delicate materials, stack machining would be an enabling process for fixturing such materials. On the other hand, stack machining would not be an option for most CNC micromachining as the miniature spindles and end mills would not be able to handle the increased load resulted from stacking.

Author Contributions: Conceptualization, H.-T.L. and N.G.; methodology, H.-T.L.; validation, H.-T.L.; formal analysis, H.-T.L.; investigation, H.-T.L.; resources, N.G.; data curation, H.-T.L.; writing—original draft preparation, H.-T.L.; writing—review and editing, N.G.; visualization, H.-T.L.; supervision, N.G.; project administration, N.G.; funding acquisition, N.G. All authors have read and agreed to the published version of the manuscript.

Funding: This research was funded partially by an OMAX IR&D project, an NSF SBIR Phase II grant number 1058278.

Acknowledgments: This work was supported by an independent research and development (IR&D) fund from OMAX Corporation. The research and development of the micro abrasive waterjet technology was supported by NSF SBIR Phase 1 and 2 Grants (No. 0944239 and 1058278). Any opinions, findings, and conclusions or recommendations expressed in this material are those of the authors and do not necessarily reflect the views of the NSF. Special thanks are to the participating technical personnel at CBA and industrial laboratories and manufacturers including Formlabs, Datron Moog Inc., and BMF Precision Technology Co, Ltd for machining/building the reference parts made from various materials and providing them for the evaluation and comparison of their performances. The authors would also like to thank Axel Henning for reviewing the paper.

Conflicts of Interest: The authors declare no conflict of interest.

Appendix A

Table A1. Performance comparison—machining single-sheet flexures under the MicroCutting Project.

Machine Tools [Nozzle Combination]	Material/ THK (mm)	Jet Diameter, Spot Size or Layer Thickness/ Particle Sizes	Position Accuracy/ Resolution	Setup Time (min)	Cut/Feed Speed (m/min)	krpm	Cutting Time (min)	Damage (Mech/ Heat)	Comments
MicroMAX [5/10] [1]	Al/0.64	φ0.3 mm/30 μm	±12 μm	~ 10	1.01	N/A	2.2 [2]	No	
MicroMAX [5/10] [1]	SS/0.51	φ0.3 mm/30 μm	±12 μm	~ 10	0.51	N/A	2.8 [2]	No	Presence of slight edge rounding and taper—materials were too thin for TAJ to remove edge taper effectively
MicroMAX [5/10] [3]	SS/0.51	φ0.3 mm/30 μm	±12 μm	~ 10	0.51	N/A	2.5 [3]	No	
MicroMAX [7/15]	Al/0.64	φ0.4 mm/60 μm	±12 μm	~ 10	1.88	N/A	2.5 [4]	No	
MicroMAX [7/15]	SS/0.75	φ0.4 mm/60 μm	±12 μm	~ 10	0.72	N/A	3.6 [4]	No	
5555 [7/15]	Al/0.80	φ0.4 mm/60 μm	±76 μm	~ 10	1.88	N/A	2.5 [4]	No	Slight edge rounding and taper—TAJ was ineffective
5555 [7/15]	SS/0.61	φ0.4 mm/60 μm	±76 μm	~ 10	0.72	N/A	2.5 [4]	No	
Zund	Al/0.51	φ0.38 mm	±0.1 mm/m	~ 15	0.07	50	≤ 8 [2]	No	Workpiece must be secured firmly to mitigate movement during milling
Zund	Al/0.51	φ 0.76 mm	±0.1 mm/m	~ 15	0.43	50	2.5 [4]	No	
Oxford laser	SS/0.61	20 μm	50 μm	~ 20	–	N/A	> 60 [4]	Some	Pulsed at 5 kHz; scrapped
FabLight fiber laser	SS/0.61	25 μm	±20 μm/m	~ 20	0.15	N/A	~ 5 [4]	Yes	Pulsed at 500 Hz; discolored with slag

Table A1. *Cont.*

Machine Tools [Nozzle Combination]	Material/ THK (mm)	Jet Diameter, Spot Size or Layer Thickness/ Particle Sizes	Position Accuracy/ Resolution	Setup Time (min)	Cut/Feed Speed (m/min)	krpm	Cutting Time (min)	Damage (Mech/ Heat)	Comments
Wire EDM	SS/0.61	ϕ0.15 mm	0.15 mm	~ 25	–	N/A	~ 60 [4]	Some	Took too long to cut. Scrapped
LPBF Moog Inc.	Al/0.51	30 µm [5]	30 µm	long	N/A	N/A	300 [4] + 30 [6]	N/A	Model EOS 290 [9]
LPBF Moog Inc.)	17-4/0.64	40 µm [5]	40 µm	long	N/A	N/A	300 [4] + 30 [6]	N/A	Model SLM 280 [10]
3D printing (DMF)	GR/0.51	10 ~ 40 µm	25 µm	Long	N/A	N/A	35 [4]	N/A	NanoArch-InP140/ InS140
3D printing (DMF)	GR/1.02	10 ~ 40 µm	25 µm	Long	N/A	N/A	70 [4]	N/A	NanoArch-InP140/ InS140
3D printing (DMF)	HTL/ 0.51	10 ~ 40 µm	25 µm	Long	N/A	N/A	35 [4]	N/A	NanoArch-InP140/ InS140
3D printing (Formlabs)	Rigid resin/ 0.85	140 µm	140 µm	Long	N/A	N/A	180 [4]	N/A	Form 2-Stereolithography (angled support)
3D printing (Formlabs)	Rigid resin/ 0.85	140 µm	140 µm	Long	N/A	N/A	240 [4]	N/A	Form 2-Stereolithography (vertical support)
Datron–Neo	Al/0.51	ϕ0.76mm [7](flexure)/ϕ3mm [8] (perimeter)	130 µm	~ 30	1.0/1.5	38/32	6.0 [2]	No	End mill diameter = gap width
Datron–Neo	Al/0.51 [2]	ϕ0.48 mm [7](flexure)/ ϕ3mm [8] (perimeter)	130 µm	~ 30	0.3/1.5	38/32	16.2 [4]	No	Gap too large with end mill diameter > gap width

[1] Nozzle combination [orifice ID/mixing tube ID in thousandth of inch]; [2] Half scale; [3] 0.4 scale; [4] Full scale; [5] Layer Thickness; [6] ϕ0.25 Wire EDM finishing time; [7] Double flute carbide end mill; [8] Single flute carbide end mill; [9] Stress relieved 300C for 3 hrs.; [10] Stress relieved 1150C for 90 min.

References

1. Liu, H.T. "7M"advantage of abrasive waterjet for machining advanced materials. *J. Manuf. Mater. Process.* **2017**, *1*, 11. [CrossRef]
2. Liu, H.T. Versatility of micro abrasive waterjet technology for machining nanomaterials. In *Dekker Encyclopedia of Nanoscience and Nanotechnology*, 3rd ed.; CRC Press: Boca Raton, FL, USA, 2017; pp. 1–18. [CrossRef]
3. Liu, H.T. Precision machining of advanced materials with abrasive waterjets. *IOP Conf. Ser. Mater. Sci. Eng.* **2017**, *164*, 012008. [CrossRef]
4. Liu, H.T. Performance comparison of waterjet on meso-micro machining of waterjet, lasers, wire EDM, and CNC milling. *Int. J. Emerg. Eng. Res. Technol.* **2019**, *7*, 31–46.
5. Wall, M. Inside NASA's Plan to Catch an Asteroid (Bruce Willis not Required). April 2013. Available online: https://www.space.com/20612-nasa-asteroid-capture-mission-explained.html (accessed on 27 February 2020).
6. Tate, K. How to Catch an Asteroid: NASA Mission Explained (Infographic). April 2013. Available online: https://www.space.com/20610-nasa-asteroid-capture-mission-infographic.html (accessed on 27 February 2020).
7. Miller, D.S. New abrasive waterjet systems to complete with lasers. In Proceedings of the 2005 WJTA Conference and Exposition, Houston, TX, USA, 21–23 August 2005.
8. Liu, H.T.; Hovanski, Y.; Caldwell, D.D.; Williford, R.E. Low-cost manufacturing of flow channels with multi-nozzle abrasive-waterjets: A feasibility investigation. In Proceedings of the 19th International Conference on Water Jetting, Nottingham, UK, 15–17 October 2008.
9. Torres, C.D.; Heaney, P.J.; Sumant, A.V.; Hamilton, M.A.; Carpick, R.W.; Pfefferkorn, F.E. Analyzing the performance of diamond-coated micro end mills. *Int. J. Mach. Tools Manuf.* **2009**, *49*, 599–619. [CrossRef]
10. Stugelmayer, E. Characterization of Process Induced Defects in Laser Powder Bed Fusion Processed Alsi10mg Alloy. Master's Thesis, Montana Tech of University of Montana, Butte, MT, USA, 2018.
11. Afkhami, S.; Dabiri, M.; Alavi, S.H.; Björk, T.; Salminen, A. Fatigue characteristics of steels manufactured by selective laser melting. *Int. J. Fatigue* **2019**, *122*, 72–83. [CrossRef]

12. Aboulkhair, N.T.; Maskery, I.; Tuck, C.; Ashcroft, I.; Everitt, N.M. Improving the fatigue behavior of a selectively laser melted aluminum alloy: Influence of heat treatment and surface quality. *Mater. Des.* **2016**, *104*, 174–182. [CrossRef]

13. Liu, H.T.; Gnäupel-Herold, T.; Hovanski, Y.; Dahl, M.E. Fatigue performance enhancement of AWJ-machined aircraft aluminum with dry-grit blasting. In Proceedings of the 2009 American Waterjet Conference, Houston, TX, USA, 18–20 August 2009; p. 15.

14. Liu, H.T.; Hovanski, Y.; Dahl, M.E. Machining of aircraft titanium with abrasive-waterjets for fatigue critical applications. *ASME J. Press. Vessel Technol.* **2012**, *134*, 011405. [CrossRef]

15. D'Urso, P.; Effeney, D.; Earwaker, W.J.; Barker, T.; Redmond, M.; Thompson, R.; Tomlinson, F. Custom cranioplasty using stereolithography and acrylic. *Br. J. Plast. Surg.* **2000**, *53*, 200–204. [CrossRef] [PubMed]

16. Ehrlich, D.; Silverman, S.; Aucoin, R.; Burns, M. Laser etching for flip-chip de-bug and inverse stereolithography for MEMS. *Solid State Technol.* **2001**, *44*, 145.

17. Liu, H.T.; Cutler, V.; Raghavan, C.; Miles, P.; Webers, N. Advanced abrasive waterjet for precision multimode machining. In *Abrasive Technology—Characteristics and Applications*; Rudawska, A., Ed.; Intech Open Access Publisher: Rijeka, Croatia, 2018; pp. 39–64. ISBN 978-953-307-906-6.

18. Liu, H.T.; McNiel, D. Versatility of waterjet technology: From macro and micro machining for most materials. In Proceedings of the 20th International Conference on Water Jetting, Graz, Austria, 20–22 October 2010.

19. Nata Rajan, Y.; Murugesan, P.K.; Mohan, M.; Khan, S.A.N.A. Abrasive water jet machining process: A state of art of review. *J. Manuf. Process.* **2020**, *49*, 271–322. [CrossRef]

20. Liu, H.T.; Schubert, E. Micro abrasive-waterjet technology (Chapter Title). In *Micromachining Techniques for Fabrication of Micro and Nano Structures*; Kahrizi, M., Ed.; Intech Open Access Publisher: London, UK, 2012; pp. 205–234. ISBN 978-953-307-906-6.

Article

Thermal Modeling of Temperature Distribution in Metal Additive Manufacturing Considering Effects of Build Layers, Latent Heat, and Temperature-Sensitivity of Material Properties

Elham Mirkoohi [1,*], Jinqiang Ning [1], Peter Bocchini [2], Omar Fergani [3], Kuo-Ning Chiang [4] and Steven Y. Liang [1]

[1] Woodruff School of Mechanical Engineering, Georgia Institute of Technology, Atlanta, GA 30332, USA; jinqiangning@gatech.edu (J.N.); steven.liang@me.gatech.edu (S.Y.L.)

[2] Carpenter Technology Corporation, Senior Additive Manufacturing Engineer, Huntsville, AL 35671, USA; pbocchini@Cartech.com

[3] Siemens Digital Factory, Product Lifecycle Management, Nonnendammallee 101, Bauteil C, 13629 Berlin, Germany; omar.fergani@siemens.com

[4] Department of Power Mechanical Engineering, National Tsing Hua University, Hsinchu 30013, Taiwan; knchiang@pme.nthu.edu.tw

* Correspondence: Elham.mirkoohi@gatech.edu

Received: 12 August 2018; Accepted: 8 September 2018; Published: 12 September 2018

Abstract: A physics-based analytical model is proposed in order to predict the temperature profile during metal additive manufacturing (AM) processes, by considering the effects of temperature history in each layer, temperature-sensitivity of material properties and latent heat. The moving heat source analysis is used in order to predict the temperature distribution inside a semi-infinite solid material. The laser thermal energy deposited into a control volume is absorbed by the material thermodynamic latent heat and conducted through the contacting solid boundaries. The analytical model takes in to account the typical multi-layer aspect of additive manufacturing processes for the first time. The modeling of the problem involving multiple layers is of great importance because the thermal interactions of successive layers affect the temperature gradients, which govern the heat transfer and thermal stress development mechanisms. The temperature profile is calculated for isotropic and homogeneous material. The proposed model can be used to predict the temperature in laser-based metal additive manufacturing configurations of either direct metal deposition or selective laser melting. A numerical analysis is also conducted to simulate the temperature profile in metal AM. These two models are compared with experimental results. The proposed model also well captured the melt pool geometry as it is compared to experimental values. In order to emphasize the importance of solving the problem considering multiple layers, the peak temperature considering the layer addition and peak temperature not considering the layer addition are compared. The results show that considering the layer addition aspect of metal additive manufacturing can help to better predict the surface temperature and melt pool geometry. An analysis is conducted to show the importance of considering the temperature sensitivity of material properties in predicting temperature. A comparison of the computational time is also provided for analytical and numerical modeling. Based on the obtained results, it appears that the proposed analytical method provides an effective and accurate method to predict the temperature in metal AM.

Keywords: metal additive manufacturing; analytical model; temperature prediction; FEA; melt pool geometry

---

1. Introduction

Metal additive manufacturing (AM) is a "process of joining materials to make objects from 3D model data, usually layer upon layer, as opposed to subtractive manufacturing methodologies" [1]. Additive manufacturing (AM) processes have potential to be the pillar of the next industrial revolution. AM can be used to improve existing manufacturing processes and rapidly introduce new prototypes and products [2,3]. It also offers the potential to spin off entirely new industries and lead to new production methods [4]. AM offers design flexibility, the ability to produce complex parts, and lower cost due to the reduced requirement of materials and decreased lead time.

AM may also hold the potential for the repair and replacement of existing plant components [5,6]. Results from on-line monitoring and complimentary non-destructive evaluation (NDE) inspections can provide indications of component health and enable repair or replacement prior to a forced outage situation. As an example, imaging tools and software can be leveraged to create a digital image of the damaged component which can be used to 3-D print a new one. This can be especially advantageous if the component is no longer in production and/or would require a long-lead time to fabricate.

There are many challenges that necessitate being focused on this field in order to expedite the adoption of AM as an advanced manufacturing technology. The issues in this field can be classified in to: the distortion, fatigue, defects, and residual stress of the manufactured parts [7–9]. The modeling in additive manufacturing technology is a key to the advancement of the field due to obstacles in in-situ measurements of temperature, thermal stress, residual stress, and distortion. The available knowledge and technology to-date on the descriptions and predictions of the metal AM process have been fragmented, mostly driven by phenomenological or numerical observations [10–13], and primarily limited to macroscopic analysis in nature [14,15], thus restricting the full capability and potential of the AM process. Using numerical methods and experiments are not just expensive, but also time-consuming. On the other hand, the physics-based analytical modeling eliminates all the above-mentioned difficulties and can help to better understand the physical aspects of the metal additive manufacturing process.

The most important part of the metal AM process modeling and prediction is the prediction of the temperature induced by laser since the non-uniform temperature will cause the thermal stress to appear in the structure. As a result of thermal stress in the build material, the tensile residual stress on the surface accelerates the crack propagation and growth [16,17]. Several researchers worked on predicting the temperature profile during the additive manufacturing process. Fergani et al. introduced an analytical model to predict the temperature in the direct metal deposition process. They predict the temperature using a moving point heat source analysis. In this work, the effect of material temperature sensitivity is ignored [18]. C.Y. Yap et al. have proposed an analytical model to predict the energy input required to process different metallic materials for selective laser melting (SLM) process. The model holds many assumptions, such as a semi-circular cross-section for melt tracks, temperature-independent specific heat, no heat loss to the surroundings and absorptance of material to laser irradiation based on bulk material properties. The melting, solidification, and solid-state phase change is also not considered in their model. The simplified model is able to predict the required energy input within an order of magnitude and provide researchers with a useful model to estimate the optimal SLM parameters [19].

Predicting the temperature precisely in metal AM is the pillar for predicting the thermal stress, residual stress, and part distortion. The non-uniform heating during AM processes may lead to the thermal stress. The large thermal gradient and cooling rate during the metal AM processes can generate complex microstructures in the build material [20]. Kelly et al. used the temperature in the AM processes in order to predict the microstructure evolution in the build part. In their work, the melting/solidification phase change is not considered [21]. Hoadley and Rappaz introduced a 2D quasi-stationary model to predict the temperature in the laser cladding process. Their research focused on the influence of the laser speed and power on the layer thickness [22]. Toyserkani et al. developed a 3D model, their proposed model tried to solve the heat problem using a coupled multi-physics system. They have used thermal analysis in order to predict the melt pool shape [23]. Cao and Ayalew have developed a control-oriented multiple input multiple output modeling of the laser-aided powder deposition processes. The objective of their work is to control the height and the temperature of a layer. Their investigation described the essential role of temperature modeling to control the quality of the final part [24]. Hitzler et al. investigated the influence of scan strategy on material characteristics, such as strength, hardness, and young's modulus [25,26]. Rashid et al. worked on the effect of scan strategy on density and metallurgical properties of a build part during the selective laser melting (SLM) process. Their results showed that parts which are made using a single scan have higher levels of hardness than parts that are made by scanning each layer twice [27].

Due to the complexity of the additive manufacturing processes, such as direct metal deposition (DMD), and SLM, not only is it time-consuming to do the experiments in order to capture the physical aspects of the metal AM processes, but it is also expensive. In the past few decades, the numerical simulations appear to be the only effective way to achieve an understanding of metal additive manufacturing processes [28,29]. The numerical methods have low computational efficiency and cannot capture all the physical aspects of the metal AM processes. On the other hand, physics-based analytical models provide a deep understanding of the physical concepts of AM. The analytical solutions have the potential to predict the key AM attributes in ways significantly faster than finite element method (FEM) simulations, by two or more orders of magnitudes [30]. Efficient and accurate predictions are therefore enabled, and the optimization of metal additive manufacturing processes which would be too complicated to cope with by the majority of other studies, who have resorted to empirical and FEM attempts. It also reduces, if not completely eliminates, the need for a costly and lengthy trial and error developmental curve for new material and components [31]. A complete build analysis with high accuracy becomes computationally tractable using the analytical model.

The AM process is a coupling of many physical phenomena such as heat transfer, fluid dynamics, phase transformation and solid mechanics. Moreover, the transient nature of heat transfer phenomena and interaction of layers make it a complicated multi-physics problem. Many researchers tried to predict the temperature in metal additive manufacturing, but each of them has several limitations. For example, not considering the temperature dependent material properties, the melting/solidification phase change, and layering aspect of metal AM. The key advantage of the proposed model is the ability to capture the most physical phenomena in metal AM, which has mostly been ignored in previous works. In this work, all the above-mentioned limitations are considered in the analytical solution of temperature. It is assumed that the thermal properties of material are temperature dependent. The melting, solidification, and solid-state phase change is included by using the modified specific heat, which relates the specific heat and latent heat of fusion. As each layer is deposited, the temperature profile is predicted using the moving heat source analysis. The laser thermal energy deposited into a control volume is absorbed by the material thermodynamic latent heat and conducted through the contacting solid boundaries. The deposited energy on the first layer introduces a thermal profile. The thermal behavior in the second pass of the laser will not be the same as the first pass since the thermal interaction of the successive layers have an influence on heat transfer. The melt pool geometry is well captured based on the proposed model, since it considers most of the previous lacks.

The outline of the paper is as follows. Section 2 presents the mathematical and practical details of the proposed analytical and numerical models. Section 3 presents detail of the experimental work which is used for validation of the proposed model, the results and a detailed discussion about the obtained results. Last but not least, Section 4 presents the conclusion of this research.

2. Approach and Methodology

2.1. Analytical Modeling

There are various engineering applications, such as turning, grinding, welding, and 3D printing in which the computation of the temperature field in the solid is modeled as a problem of heat conduction involving a moving heat source. The objective of this section is to present the mathematical formulation and the method of solution of heat conduction by considering the moving heat source, which indeed is the case in metal additive manufacturing [32].

In this study, the basic premise is that the powder is situated in the desirable location relative to the melt pool. In other words, there is no moment or mass transfer consideration in this work, and only the heat transfer is considered. Although the effect of time difference between the two consecutive irradiations on temperature profile is not considered in this work, it is worth noting that considering the existence of the time difference between two consecutive irradiations may cause an increase in predicted temperature during the metal AM processes. This is because the predicted temperature at time $t + \Delta t$ will be the materials-response-coupled superposition considering the temperature sensitivity of thermal properties at time t and $t + \Delta t$.

By considering a line heat source of constant strength g_l^c (W/m) located at the x-axis and oriented parallel to the z-axis, the source releases its energy continuously over time as it moves with a constant velocity of v in the positive x-direction. The medium is initially at room temperature. It is assumed $(\partial T/\partial z) = 0$ everywhere in the medium. Hence, the differential equation of heat conduction in the x, y coordinates in now taken as

$$\frac{\partial^2 T}{\partial x^2} + \frac{\partial^2 T}{\partial y^2} + \frac{1}{k}g(x,y,t) = \frac{1}{\alpha}\frac{\partial T}{\partial t} \tag{1}$$

where $T \equiv T(x,y,t)$. k is thermal conductivity, and α is the thermal diffusivity. The line heat source g_l^c (W/m) is related to the equivalent volumetric source $g(x,y,t)$ (W/m^3) by the delta function notation as

$$g(x,y,t) = g_l^c\,\delta(y)\delta(x - vt) \tag{2}$$

In order to consider the moving heat source, it is assumed that the coordinate system transfers from the x, y fixed coordinate system to ζ, y coordinate moving with the line heat source by using the transformation

$$\zeta = x - vt \tag{3}$$

Using the abovementioned transformation, the heat conduction equation for the moving coordinate system can be written as

$$\frac{\partial^2 T}{\partial \zeta^2} + \frac{\partial^2 T}{\partial y^2} + \frac{1}{k}g_l^c\delta(\zeta)\delta(y) = \frac{1}{\alpha}\left(\frac{\partial T}{\partial t} - v\frac{\partial T}{\partial \zeta}\right) \tag{4}$$

Equation (4) can be solved by the assumption of the quasi-stationary condition [32]. Using the separation of variables, the closed form solution of the temperature field can be obtained as

$$T = \frac{P}{4\pi KR}\exp\frac{-v(R - x)}{2\alpha} + T_0 \tag{5}$$

where P is the laser power, K is thermal conductivity, v is scan speed (laser velocity). α is thermal diffusivity which can be calculated as $\frac{K}{\rho c}$. In which ρ is material density and c is material heat capacity. $R = \sqrt{x^2 + y^2}$ is the radial distance from the heat source. T_0 is the initial temperature. The material is considered homogeneous and isotropic. As the laser moves along the surface it deposited some energy. Figure 1 depicts the heat transfer in metal AM. The heat loss from the surface by radiation and convection are not considered in this study.

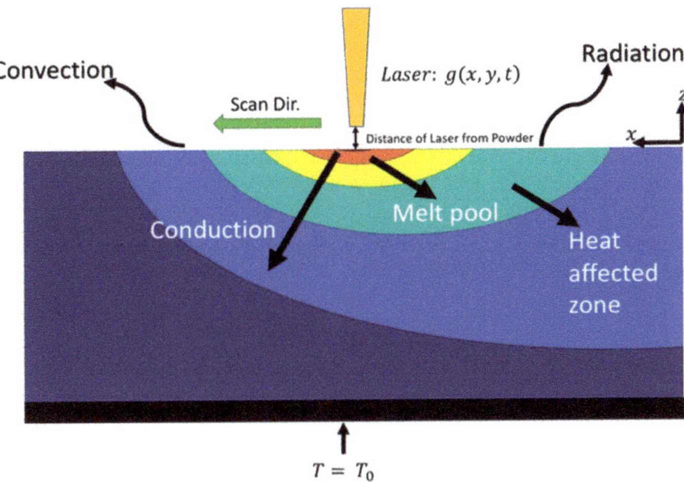

Figure 1. Heat transfer during laser-based metal additive manufacturing.

It is worth noting that the process parameters such as laser power, scanning speed, powder size, powder distribution, etc. have influence on material properties in the metal AM processes since it may change the predicted temperature [33]. As a result, the material properties are assumed to be temperature dependent as shown in Table 1 [34].

Table 1. Material properties of Ti-6Al-4V.

Properties	Ti-6Al-4V
Liquidus temperature (K)	1928
Solidus temperature (K)	1878
Thermal conductivity (W/m K)	$\begin{cases} K_s = 1.57 + 1.6 \times 10^{-2}T - 1 \times 10^{-6}T^2 & 1268 < T < 1928 \\ K_l = 33.4 & T = 1928 \\ K_l = 34.6 & T = 1978 \end{cases}$
Specific heat (J/Kg K)	$\begin{cases} C_p = 492.4 + 0.025T - 4.18 \times 10^{-6}T^2 & 1268 < T < 1928 \\ C_p = 830 & T > 1928 \end{cases}$
Density (Kg/m^3)	4420
Viscosity (Kg/m s)	4×10^{-3}
Latent heat (J/Kg)	2×10^5

During the metal AM process such as SLM and DMD, the melting, solidification and solid-state phase transformation take place. This is considered using modified heat capacity.

$$C_P^m = C_P(T) + L_f \frac{\partial f}{\partial T} \tag{6}$$

In which $C_p(T)$ is temperature dependent specific heat, L_f is latent heat of fusion, and f is liquid fraction which can be calculated from

$$f = \begin{cases} 0, & T < T_s \\ \frac{T - T_s}{T_L - T_s}, & T_s < T < T_L \\ 1, & T > T_L \end{cases} \tag{7}$$

where, T_s is solidus temperature and T_L is liquidus temperature.

The process parameters such as laser power and scan speed are defined to start the calculation of the vertical distribution of temperature during the laser-based metal AM. At first, it is assumed the powder is at room temperature. As the laser moves along the *x*-axis, it deposits the energy on the powder and causes the powder to melt, as the laser passes the affected region, the melt pool starts to solidify. As it creates the first layer, the temperature profile is calculated for that layer. Next, it starts the second layer with the dwell time of zero. It is possible that the first layer has not had enough time to cool down to the room temperature when the second layer is starting to build. As a result, it affects the heat transfer during the metal AM processes. Considering the layer addition also has a substantial influence on thermal stress and residual stress predictions. The fellow chart of considering the build layers is illustrated in Figure 2.

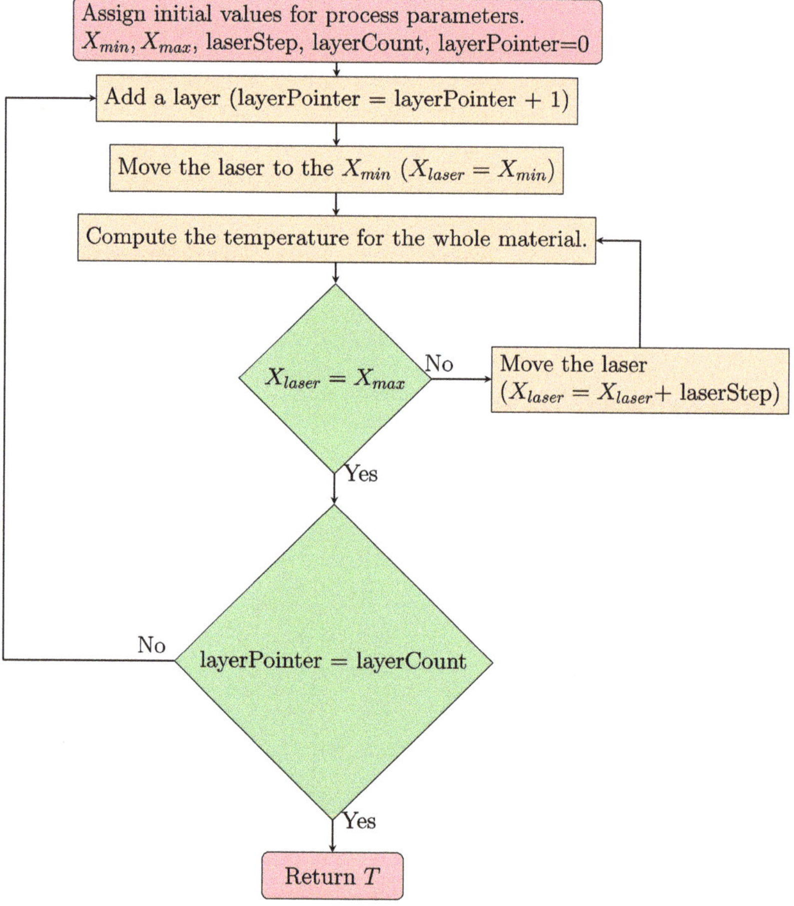

Figure 2. Fellow chart of considering build layers.

2.2. Numerical Modeling

For further validation of this work, finite element analysis is used. The temperature profile is modeled using a moving heat source analysis. The user defined functions (UDF) code is written in ANSYS Fluent software using Equation (8) in order to run a FEA on a 2D geometry, as shown in Figure 3. The build part material is Ti-6Al-4V. The heat loss from the surface due to conduction and radiation is considered. The material properties are assumed to be temperature dependent as shown in Figure 4.

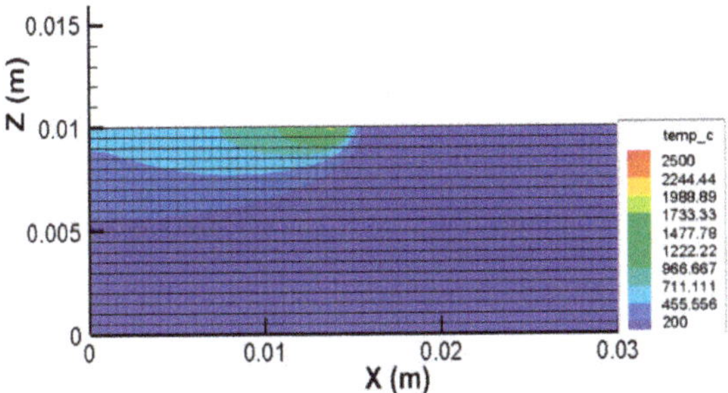

Figure 3. Representation of the mesh and numerical model.

The geometry of the build part is a rectangle shape of 30 × 10 mm. The quadratic element with the mesh size of 0.5 mm is chosen for all the simulations, as shown in Figure 3.

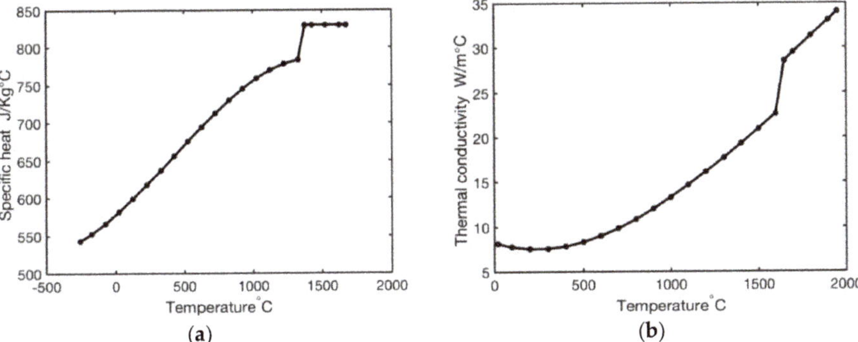

Figure 4. Material properties as a function of temperature, (**a**) specific heat, (**b**) thermal conductivity.

The laser power distribution on the laser beam focus plane is described by the Gaussian equation [32], as

$$q(x,y) = D\,\frac{P}{\pi r^2}e^{\left(\frac{-B(x-vt)^2}{r^2}\right)} \tag{8}$$

where, P is the total laser power input, r is laser spot radius, v is scanning speed B is gaussian shape factor, and D is a numerical parameter used to fit the experimental data. It accounts for the absorptivity of the material, the heat lost to the metal powder before it falls into the melt pool and the angle of the surface with the laser beam. The values of the process parameters are listed in Table 2. The melting temperature of Ti-6Al-4V is in the range of 1538–1649 °C. In this study, 1620 is selected as the melting temperature of Ti-6Al-4V. The two-dimensional heat transfer in a rectangular surface could be described by

$$\rho C\left(\frac{\partial T}{\partial t} + v\frac{\partial T}{\partial x}\right) = \nabla(k\nabla T) + S \tag{9}$$

where ρ is material density, C is specific heat, k is thermal conductivity, and S is the heat sink.

The boundary condition on the laser heating surface is defined as

$$k\frac{\partial T}{\partial y} = q(x,y) - h(T - T_0) - \sigma\varepsilon(T^4 - T_0^4) \tag{10}$$

where $q(x,y)$ is laser power input, h is the heat transfer coefficient, σ is the thermal radiation coefficient, ε is the material emissivity, T_0 is the ambient temperature. The initial condition could be as

Table 2. Material parameters used for numerical modeling.

Name	Value
Thermal radiation coefficient (W/m$^2\cdot{}^\circ$C^4)	5.67×10^{-8}
Heat transfer coefficient (W/m$^2\cdot{}^\circ$C)	24
Material emissivity	0.9
D	[0.2–0.4]
Gaussian shape factor	2
Laser spot radius (mm)	0.7
Ambient temperature	25

3. Modeling Results and Experimental Comparison

3.1. Temperature Profile, Maximum Temperature, and Surface Temperature

In this section, the temperature profile, maximum temperature and surface temperature are predicted and compared to the experimental results. A moving heat source analysis is used in order to predict the temperature distribution associated with the dynamic heat deposition. The explicit and closed-form temperature solutions are calculated in Section 2.1. The general differential equation of heat conduction in the 2D plane is used. In order to consider the moving heat source, it is assumed that the coordinate system moves with the heat source by using a transformation as shown in Equation (3). Finally, using the separation of variables, the closed-form solution of temperature is obtained in Equation (5). The material properties are assumed to be temperature dependent. The melting/solidification phase change is also considered. The analytical and numerical analysis are conducted in this work.

In order to validate the proposed model, the experimental temperature data are used from the work of Pauzet [35]. The Ti-6Al-4V samples are manufactured using the DMD machine. The dimensions of the samples are 2 mm in width, 70 mm in depth and 80 mm in length. The temperature on the build part surface is measured using the thermocouple of type K. In order to control the experimental setup, the authors used a thermal-camera and a high-speed camera to provide comparison bases for the temperature and the melt-pool size. The DMD machine has used the laser with the wavelength of 1030 nm. The scanning speed of 0.2 m/min and 0.4 m/min and the laser power of 400 W and 600 W are studied. The initial temperature of each layer depends on the final temperature of the previous layer, as the process is multi-layered.

Figure 5 shows the temperature profile of the build part. The temperature is predicted using both the analytical model and the numerical model. The laser moves along the x-axis from left to right. The small red spot on top shows the laser location. The layer thickness is chosen to be 80 μm. The distance of the laser from the powder is 0.4 mm. For the same power, as the velocity is increased the maximum temperature is decreased since the powder has less time to absorb the energy. Different combinations of the process parameters are presented in Figure 5, specifically scanning speed and laser power.

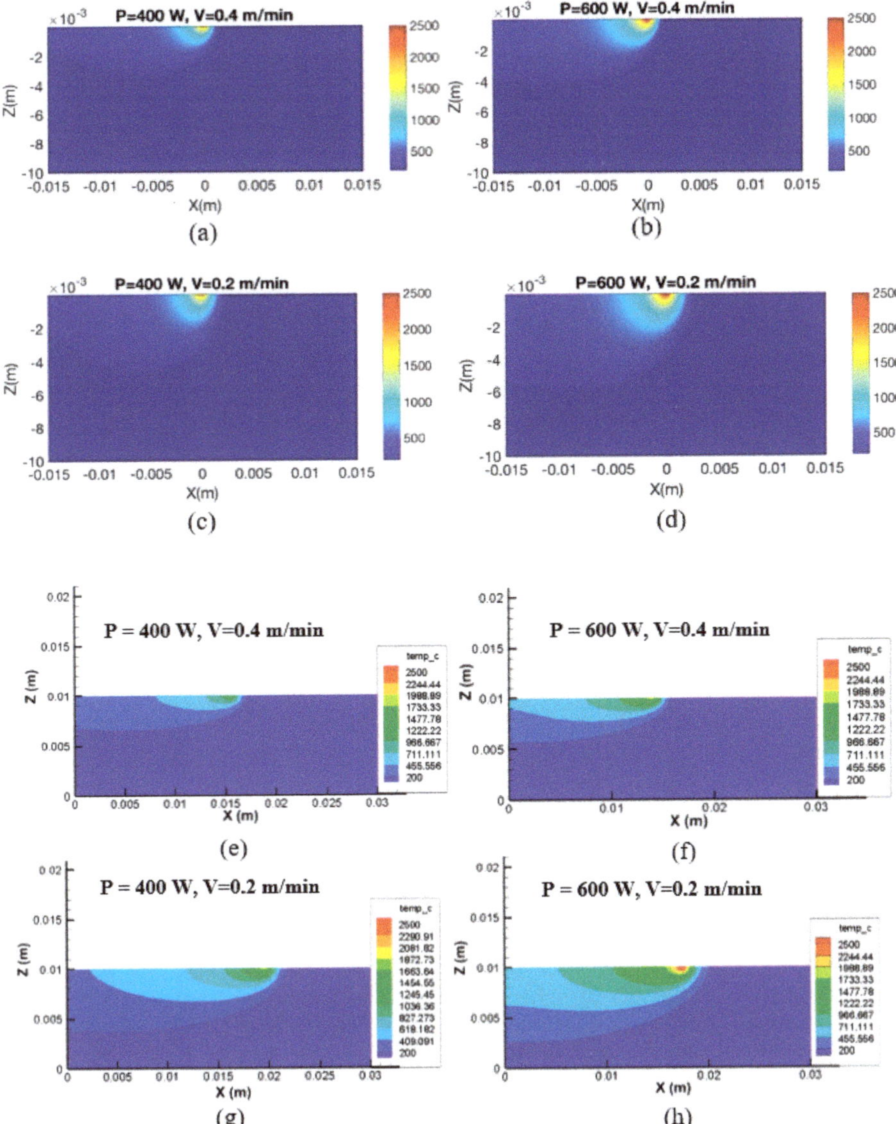

Figure 5. Predicted temperature profile using (**a–d**) physics-based modeling and (**e–h**) numerical modeling.

The evolution of the surface temperature is plotted as a function of time for each case as shown in Figure 6. A study point will be chosen from the 2D geometry. When the laser is far away from the study point, the powder is at room temperature. As the laser approaches the study point, the temperature increases continuously. The maximum temperature on the curve corresponds to the moment that the laser is above the study point. After the laser passes the point, the temperature is decreased which shows that the material is cooling down. As shown in these plots, the cooling rate in the AM process is substantially high.

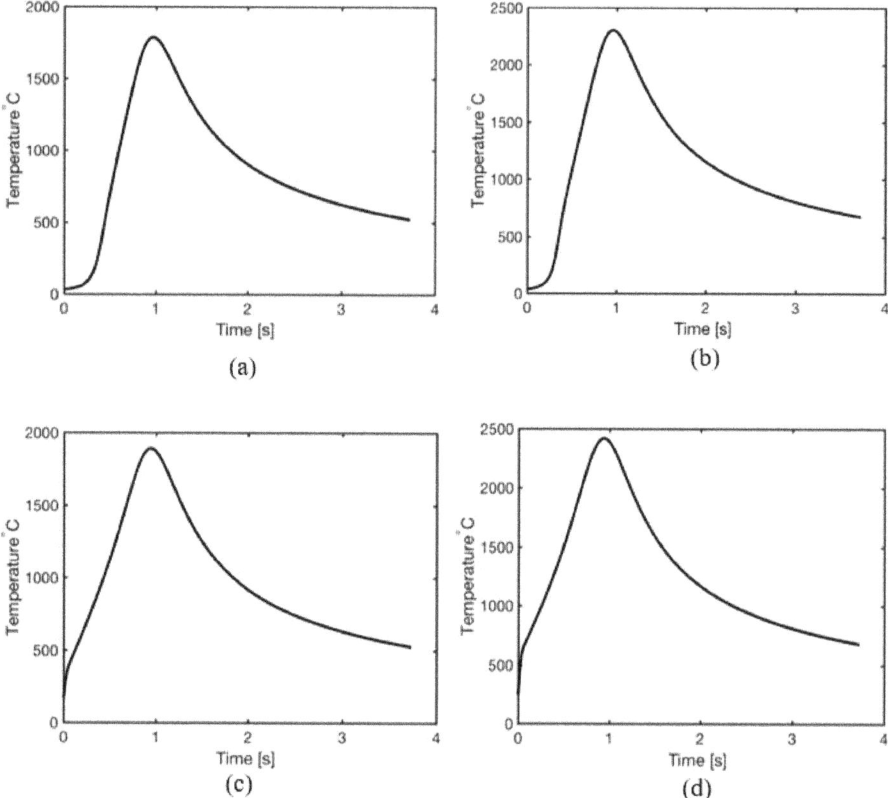

Figure 6. Evolution of surface temperature as a function of time for (**a**) P = 400 W, V = 0.4 m/min, (**b**) P = 600 W, V = 0.4 m/min, (**c**) P = 400 W, V = 0.2 m/min, (**d**) P = 600 W, V = 0.2 m/min.

In order to understand the influence of the process parameters on the maximum temperature, and surface temperature, a sensitivity study is designed to investigate both the scan speed and laser power. The short computational time associated with the analytical modeling approach allows for a better understanding of the influence of the process parameters as discussed previously. Figure 7 depicts the influence of the scan speed and laser power on temperature, as predicted by the analytical model and compared to the experimental results.

The results of the simulations from the analytical model illustrates that the maximum temperature decreases linearly as the scan speed increases since the material has less time to absorb the energy. On the other hand, for the fixed scanning speed, as the power increases the maximum temperature increases. The four experimental data are also pointed in Figure 7. The predicted temperature from the analytical model is slightly higher than the experimental values. This error is mainly because the temperature is measured using thermocouples which are a little below the surface.

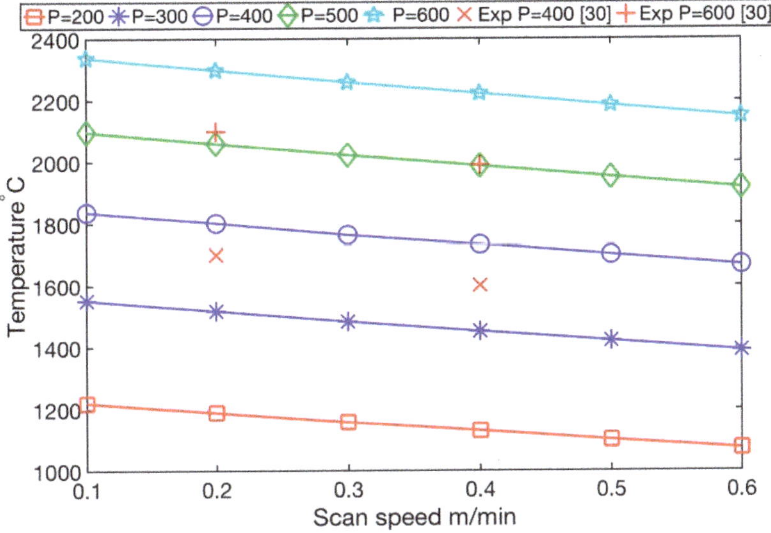

Figure 7. Effect of scan speed and laser power on peak temperature.

Figure 8 represents the influence of the laser on the surface temperature. As the power increases from 200 W to 600 W, the surface temperature increases for a fix scanning speed. On the other hand, the surface temperature will decrease as the scanning velocity increases from 0.1 m/min to 0.6 m/min for a fix laser power as shown in Figure 9.

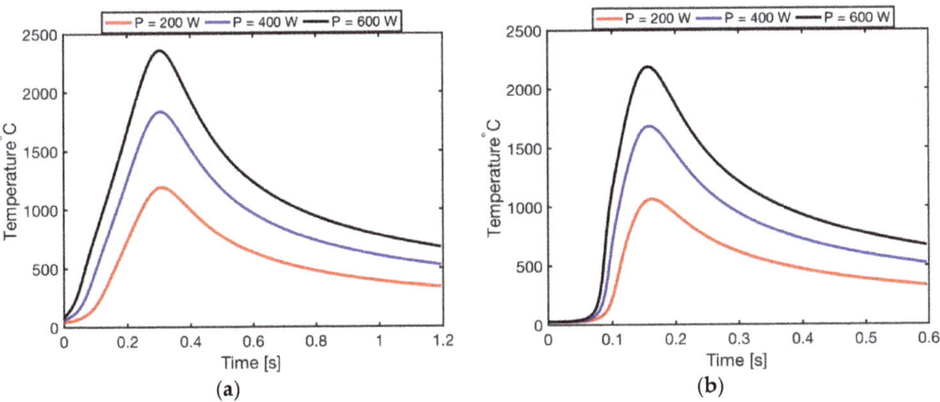

Figure 8. Comparison of evolution of surface temperature for (**a**) V = 0.3 m/min, and (**b**) V = 0.6 m/min.

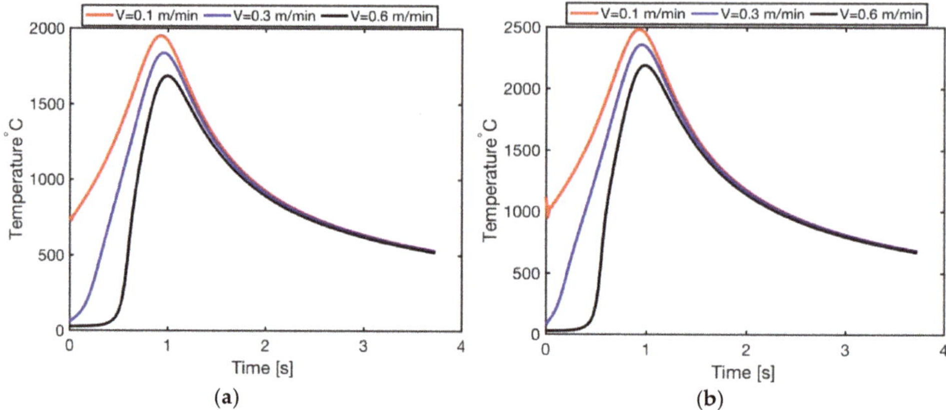

Figure 9. Comparison of evolution of surface temperature for (**a**) P = 400 W, and (**b**) P = 600 W.

As explained, the proposed model considers the multi-layer aspects of metal additive manufacturing. The effect of considering the layer addition on peak temperature is compared to the obtained peak temperature without considering the layer addition, and also compared to the experimental results.

To further validate the proposed model, the peak temperature is plotted as a function of scanning speed for different laser powers. Two different values of laser power (400 W and 600 W) and scanning speed (0.2 m/min and 0.4 m/min) are chosen. The temperature considering the layer addition, the temperature not considering the layer addition, and also experimental values are compared. The values are listed in Table 3. The observations show that considering layer addition improves the prediction of temperature, as shown in Figure 10. For example, the predicted temperature for scanning velocity of 0.2 m/min and laser power of 400 W without considering the layer addition is 2042 $^{\circ}$C, but when considering the layering aspect of AM, the predicted temperature reduces to 1802.8 $^{\circ}$C which shows that it affects the heat transfer mechanisms.

Table 3. Comparison of temperature prediction among considering layer addition, not considering the layer addition, and experimental values.

Laser Power (W)	400	400	600	600
Scanning Speed (m/min)	0.2	0.4	0.2	0.4
Max Temperature w/o Layer	2043.7	1998.1	2603.7	2538.1
Max Temperature with Layer	1802.8	1733.6	2298.1	2222.7
Experimental Values	1730	1605	2100	1970

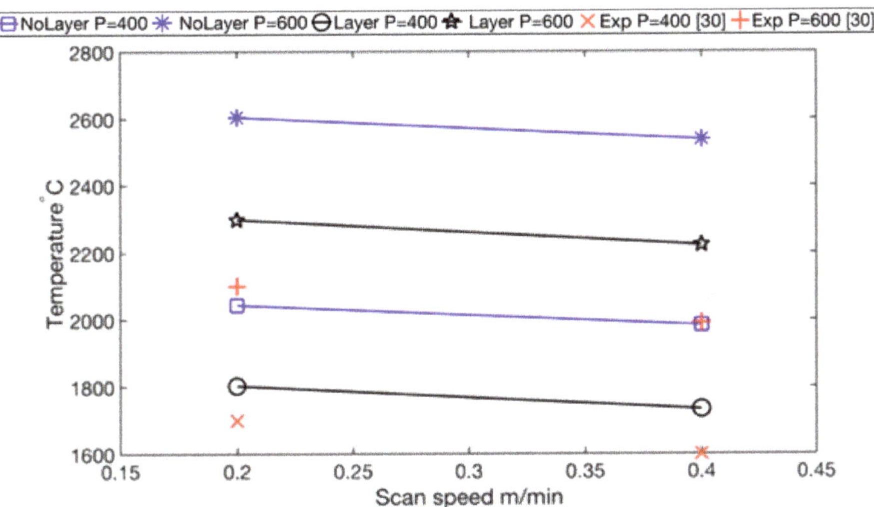

Figure 10. Comparison of prediction of temperature with and without considering the layers with experimental values.

A comparison is also conducted among the analytical model considering the layer addition and dwell time, numerical model and experimental values as shown in Figure 11.

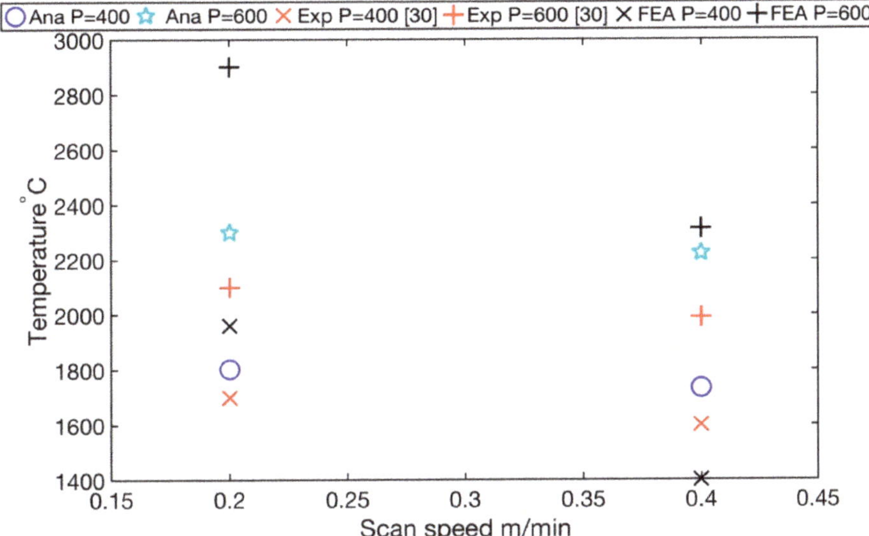

Figure 11. Comparison of predicted temperature among analytical model, experimental values, and FEA.

Overall, the temperature on the surface in terms of magnitude is well captured by both analytical and numerical approaches. The analytical model better approached the experimental measurements. This comparison shows the capability to accurately predict the temperature profile on the surface using the analytical modeling. The analytical approach also provides the power of a short computational time.

In order to illustrate the importance of considering the temperature dependent material properties, a sensitivity analysis is conducted to compare the predicted surface temperature with and without considering the property's temperature-sensitivity. The obtained results demonstrate a significant difference between them as shown in Figure 12. The thermal conductivity of the Ti-6Al-4V is 6.7 W/m·°C which results in a low rate of heat transfer in the build part. However, the thermal conductivity of Ti-6Al-4V varies from 6 to 35 W/m·°C with respect to temperature. The increase in heat transfer rate induced by the increase in thermal conductivity, causes the predicted surface temperature decrease. In the cases that the temperature sensitivity of the material properties is considered, as the velocity increases from 0.2 m/min to 0.4 m/min, the variation of predicted surface temperature is less than 100 °C. However, when the temperature sensitivity of material properties is not considered, the variation of temperature is more than 1000 °C. As it is shown in Figure 12 the predicted temperature can be quite unrealistic without considering the material properties sensitivity to temperature.

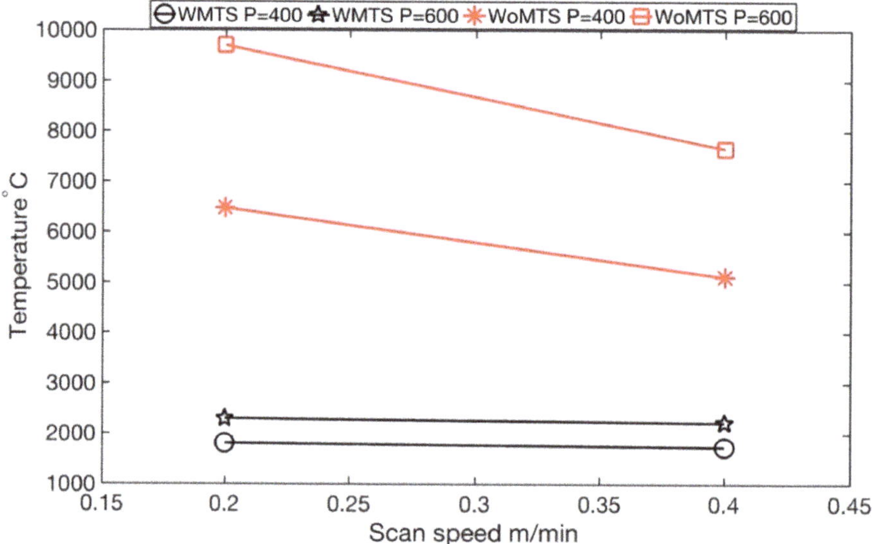

Figure 12. Comparison of predicted temperature considering the temperature sensitivity of material properties (WMTS), and without temperature sensitivity of material properties (WoMTS).

3.2. Experimental Validation Based on Melt Pool Geometry Measurement

In order to predict the morphology of the manufactured part, most of the researchers have used FEA or empirical models [36,37]. The proposed analytical model is used to predict the melt pool size. A comparison between the model and experimental results are conducted. The different process parameters such as laser power and scanning speed are used to predict the melt pool geometry. Figure 13 shows the experimental measurement of melt pool size from Peyre [38]. In this experiment, a high-speed C-Mos camera (Fastcam Photron) is used to measure the melt pool size which is generated by the DMD process.

Figure 14 demonstrates the predicted melt pool size and geometry for different process parameters in metal AM. The laser distance from powder is 1 μm.

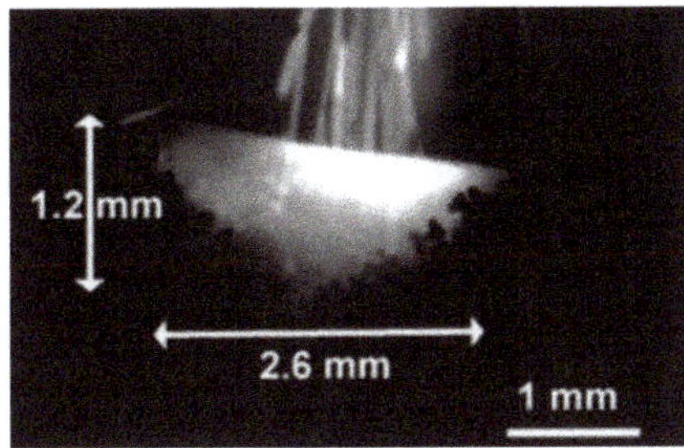

Figure 13. Experimental measurement of melt pool size for P = 600 W and V = 6 mm/s.

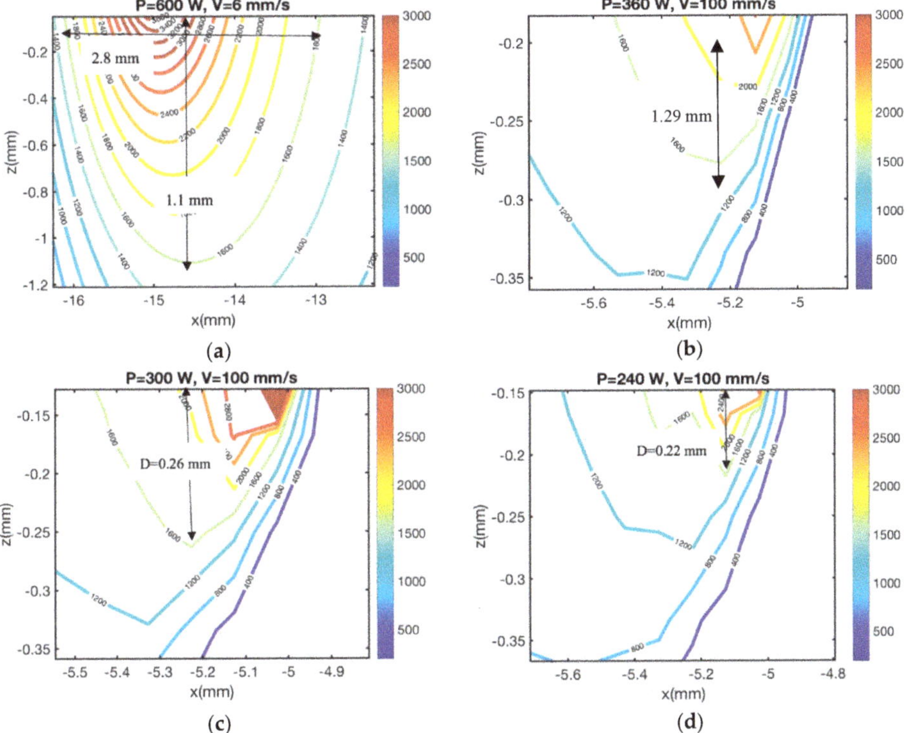

Figure 14. Predicted melt pool size in metal AM process for (**a**) P = 600 W, V = 6 mm/s (**b**) P = 360 W, V = 100 mm/s (**c**) P = 300 W, V = 100 mm/s, (**d**) P = 240 W, V = 100 mm/s

As shown in Figure 14, the melt pool depth and length are obtained using the analytical solution of temperature that is given in Section 2.1. The maximum error in length and depth is 7.6% and 3.7%, respectively. Table 4 listed the process parameters, predicted melt pool size, the experimental values, and also the corresponding error. Based on the calculated error, it is shown that the proposed 2D model can accurately capture the melt pool size. As a result, it eliminates the needs for doing costly experiments and also time-consuming FEM.

Table 4. Predicted and experimental measurements of melt pool size.

P (W)	V (mm/s)	Melt Pool Length (mm) Model	Melt Pool Length (mm) Exp/sim	Melt Pool Depth (mm) Model	Melt Pool Depth (mm) Exp/sim	Error in Length	Error in Depth
600 [38]	6	2.80	2.60	1.10	1.20	7.60%	2.00%
360 [37]	100	-	-	0.29	0.30	-	3.40%
300 [37]	100	-	-	0.26	0.27	-	3.70%
240 [37]	100	-	-	0.22	0.20	-	1.00%

4. Conclusions

Analytical models and numerical model are used to predict the temperature in laser-based metal additive manufacturing configurations of either direct metal deposition or selective laser melting. In the past few decades, many researchers have been trying to understand the relationships between the process parameters and temperature using FEM. The numerical methods have low computational efficiency, and it cannot capture all the physical aspects of the metal AM processes. The lack of a physics-based analytical model that captures all the physical phenomena of the AM processes is sensible. The physics-based analytical modeling provides accurate results. The high computational efficiency and easy implementation are the other advantages of the analytical model for the additive manufacturing modeling.

In this work, an analytical model is proposed to predict the distribution of the temperature profile by considering the interaction of the layers during the laser metal additive manufacturing process. The material properties are assumed to be temperature dependent, and also the melting/solidification phase change is considered in this work. The temperature profile, the peak temperature, and the evolution of surface temperature are obtained from the proposed model. The analytical model of the temperature is based on the moving heat source assumption, as described in Section 2. The general differential equation of heat conduction is used to obtain the closed-form temperature solution using the separation of variables in a semi-infinite medium. The material is assumed to be homogeneous and isotropic. The predicted temperature from the analytical model are compared with the experimental values and FEM results. For further validation, a comparison of peak temperature considering the layer addition and without considering the layer addition is conducted and compared with experimental values.

The results of the temperature distribution considering the layering aspects of metal additive manufacturing showed better agreement with experimental values in comparison with the predicted temperature not considering the layer addition. The observations suggest that for a fixed laser power, the laser speed increases as the temperature decreases, since the material has less time to absorb the energy. Also, for a given scanning speed, the laser power increases as the maximum temperature increases.

A numerical model is also used to predict the temperature in the metal additive manufacturing process. The material properties are assumed to be temperature dependent. In the numerical model, the heat loss due to convection and radiation is considered. The temperature is well obtained using numerical models.

A comparison is conducted in order to capture the effect of considering the temperature sensitivity of material properties. A significant difference is observed between them. The main reason is that the thermal conductivity of Ti-6Al-4V is low, so the heat transfer rate decreases and causes the surface temperature to increase. However, by considering the temperature dependent material properties, the thermal conductivity increases by increasing the temperature. As a result, the heat transfer rate increases and causes the obtained surface temperature to be less compared to the case in which the thermal conductivity is constant.

The proposed model can also predict the melt pool size with the error margin being less than 7.6%. Hence, it eliminates the costly experiments and time-consuming FEM for predicting the melt pool size. This 2D model also shows that there is no need for doing 3D simulations in order to predict the melt pool size and geometry.

The proposed analytical model shows a good agreement with the experimental values. The proposed analytical model reduces the computational time to a fraction when compared to finite element analysis. The analytical model has also eliminated the costly experiments in order to understand the physical concepts of laser metal additive manufacturing. The influence of scanning speed and laser power on the temperature profile, surface temperature, and also peak temperature are investigated and the relations between them are established.

Author Contributions: E.M. conceived and developed the proposed analytical model, extracted and analyzed the data, and wrote the paper. J.N., P.B., O.F., and K.-N.C. provided general guidance. S.Y.L. provided general guidance and proofread the manuscript writing.

Funding: The funding is confidential.

Conflicts of Interest: The authors declare no conflict of interest.

References

1. Mellor, S.; Hao, L.; Zhang, D. Additive manufacturing: A framework for implementation. *Int. J. Prod. Econ.* **2014**, *149*, 194–201. [CrossRef]
2. Bikas, H.; Stavropoulos, P.; Chryssolouris, G. Additive manufacturing methods and modelling approaches: A critical review. *Int. J. Adv. Manuf. Technol.* **2016**, *83*, 389–405. [CrossRef]
3. Turner, B.N.; Strong, R.; Gold, S.A. A review of melt extrusion additive manufacturing processes: I. Process design and modeling. *Rapid Prototyp. J.* **2014**, *20*, 192–204. [CrossRef]
4. Kruth, J.-P.; Levy, G.; Klocke, F.; Childs, T. Consolidation phenomena in laser and powder-bed based layered manufacturing. *CIRP Ann.* **2007**, *56*, 730–759. [CrossRef]
5. Woesz, A. Rapid prototyping to produce porous scaffolds with controlled architecture for possible use in bone tissue engineering. In *Virtual Prototyping & Bio Manufacturing in Medical Applications*; Springer: New York, NY, USA, 2008; pp. 171–206.
6. Frazier, W.E. Metal additive manufacturing: A review. *J. Mater. Eng. Perform.* **2014**, *23*, 1917–1928. [CrossRef]
7. Vasinonta, A.; Beuth, J.L.; Griffith, M.L. Process maps for controlling residual stress and melt pool size in laser-based SFF processes. In Proceedings of the Solid Freeform Fabrication Symposium, Austin, TX, USA, 7–9 August 2000; p. 206.
8. Pang, T.H.; Guertin, M.D.; Nguyen, H.D. Accuracy of Stereolithography Parts: Mechanism and Modes of Distortion for a "Letter H" Diagnostic Part. In Proceedings of the Solid Free Form Fabrication Symposium, Austin, TX, USA, 7–9 August 1995; pp. 170–180.
9. Zhang, Y.; Chou, Y.K. A parametric study of part distortions in FDM using 3D FEA. In Proceedings of the 17th Solid Freeform Fabrication Symposium, Austin, TX, USA, 14–16 August 2006; pp. 410–420.
10. Roberts, I.A.; Wang, C.; Esterlein, R.; Stanford, M.; Mynors, D. A three-dimensional finite element analysis of the temperature field during laser melting of metal powders in additive layer manufacturing. *Int. J. Mach. Tools Manuf.* **2009**, *49*, 916–923. [CrossRef]
11. Michaleris, P. Modeling metal deposition in heat transfer analyses of additive manufacturing processes. *Finite Elem. Anal. Des.* **2014**, *86*, 51–60. [CrossRef]
12. Krol, T.; Seidel, C.; Zaeh, M. Prioritization of process parameters for an efficient optimisation of additive manufacturing by means of a finite element method. *Procedia CIRP* **2013**, *12*, 169–174. [CrossRef]
13. Hu, D.; Kovacevic, R. Sensing, modeling and control for laser-based additive manufacturing. *Int. J. Mach. Tools Manuf.* **2003**, *43*, 51–60. [CrossRef]
14. Contuzzi, N.; Campanelli, S.; Ludovico, A. 3 D Finite Element Analysis in the Selective Laser Melting Process. *Int. J. Simul. Model.* **2011**, *10*, 113–121. [CrossRef]
15. Ding, J.; Colegrove, P.; Mehnen, J.; Ganguly, S.; Almeida, P.S.; Wang, F.; Williams, S. Thermo-mechanical analysis of Wire and Arc Additive Layer Manufacturing process on large multi-layer parts. *Comput. Mater. Sci.* **2011**, *50*, 3315–3322. [CrossRef]

16. Denlinger, E.R.; Heigel, J.C.; Michaleris, P.; Palmer, T. Effect of inter-layer dwell time on distortion and residual stress in additive manufacturing of titanium and nickel alloys. *J. Mater. Process. Technol.* **2015**, *215*, 123–131. [CrossRef]

17. Aggarangsi, P.; Beuth, J.L. Localized preheating approaches for reducing residual stress in additive manufacturing. In Proceedings of the SFF Symposium, Austin, TX, USA, 14–16 August 2006; pp. 709–720.

18. De La Batut, B.; Fergani, O.; Brotan, V.; Bambach, M.; El Mansouri, M. Analytical and Numerical Temperature Prediction in Direct Metal Deposition of Ti6Al4V. *J. Manuf. Mater. Process.* **2017**, *1*, 3. [CrossRef]

19. Yap, C.Y.; Tan, H.K.; Du, Z.; Chua, C.K.; Dong, Z. Selective laser melting of nickel powder. *Rapid Prototyp. J.* **2017**, *23*, 750–757. [CrossRef]

20. Galarraga, H.; Warren, R.J.; Lados, D.A.; Dehoff, R.R.; Kirka, M.M.; Nandwana, P. Effects of heat treatments on microstructure and properties of Ti-6Al-4V ELI alloy fabricated by electron beam melting (EBM). *Mater. Sci. Eng. A* **2017**, *685*, 417–428. [CrossRef]

21. Kelly, S.; Kampe, S. Microstructural evolution in laser-deposited multilayer Ti-6Al-4V builds: Part II. Thermal modeling. *Metall. Mater. Trans. A* **2004**, *35*, 1869–1879. [CrossRef]

22. Hoadley, A.; Rappaz, M. A thermal model of laser cladding by powder injection. *Metall. Trans. B* **1992**, *23*, 631–642. [CrossRef]

23. Toyserkani, E.; Khajepour, A.; Corbin, S. 3-D finite element modeling of laser cladding by powder injection: Effects of laser pulse shaping on the process. *Opt. Lasers Eng.* **2004**, *41*, 849–867. [CrossRef]

24. Cao, X.; Ayalew, B. Control-oriented MIMO modeling of laser-aided powder deposition processes. In Proceedings of the American Control Conference (ACC), Chicago, IL, USA, 1–3 July 2015; IEEE: Piscataway, NJ, USA, 2015; pp. 3637–3642.

25. Hitzler, L.; Hirsch, J.; Heine, B.; Merkel, M.; Hall, W.; Öchsner, A. On the anisotropic mechanical properties of selective laser-melted stainless steel. *Materials* **2017**, *10*, 1136. [CrossRef] [PubMed]

26. Hitzler, L.; Merkel, M.; Hall, W.; Öchsner, A. A Review of Metal Fabricated with Laser-and Powder-Bed Based Additive Manufacturing Techniques: Process, Nomenclature, Materials, Achievable Properties, and its Utilization in the Medical Sector. *Adv. Eng. Mater.* **2018**, *20*, 1700658. [CrossRef]

27. Rashid, R.; Masood, S.; Ruan, D.; Palanisamy, S.; Rashid, R.R.; Brandt, M. Effect of scan strategy on density and metallurgical properties of 17-4PH parts printed by Selective Laser Melting (SLM). *J. Mater. Process. Technol.* **2017**, *249*, 502–511. [CrossRef]

28. Wang, L.; Jiang, X.; Zhu, Y.; Zhu, X.; Sun, J.; Yan, B. An approach to predict the residual stress and distortion during the selective laser melting of AlSi10Mg parts. *Int. J. Adv. Manuf. Technol.* **2018**, *97*, 3535–3546. [CrossRef]

29. Shiva, S.; Palani, I.; Paul, C.; Singh, B. Laser annealing of laser additive–manufactured Ni-Ti structures: An experimental–numerical investigation. *Proc. Inst. Mech. Eng. Part B J. Eng. Manuf.* **2018**, *232*, 1054–1067. [CrossRef]

30. Mirkoohi, E.; Bocchini, P.; Liang, S.Y. An Analytical Modeling for Designing the Process Parameters for Temperature Specifications in Machining. *Preprints* **2018**. [CrossRef]

31. Mirkoohi, E.; Bocchini, P.; Liang, S.Y. An analytical modeling for process parameter planning in the machining of Ti-6Al-4V for force specifications using an inverse analysis. *Int. J. Adv. Manuf. Technol.* **2018**, *98*, 2347–2355. [CrossRef]

32. Carslaw, H.; Jaeger, J. *Conduction of Heat in Solids: Oxford Science Publications*; Oxford University Press: Oxford, UK, 1959.

33. Mirkoohi, E.; Malhotra, R. Effect of Particle Shape on Neck Growth and Shrinkage of Nanoparticles. In Proceedings of the ASME 2017 12th International Manufacturing Science and Engineering Conference Collocated with the JSME/ASME 2017 6th International Conference on Materials and Processing, Los Angeles, CA, USA, 4–8 June 2017; American Society of Mechanical Engineers: New York, NY, USA, 2017; p. V002T001A026.

34. Mills, K.C. *Recommended Values of Thermophysical Properties for Selected Commercial Alloys*; Woodhead Publishing: Cambridge, UK, 2002.

35. Pouzet, S.B. *Fabrication Additive de Composites à Matrice titane par Fusion Laser de Poudre Projetée*; ENSAM: Paris, France, 2015.

36. Pinkerton, A.J.; Li, L. Modelling the geometry of a moving laser melt pool and deposition track via energy and mass balances. *J. Phys. D Appl. Phys.* **2004**, *37*, 1885–1895. [CrossRef]

37. Cheng, B.; Chou, K. Melt pool geometry simulations for powder-based electron beam additive manufacturing. In Proceedings of the 24th Annual International Solid Freeform Fabrication Symposium—An Additive Manufacturing Conference, Austin, TX, USA, 12–14 August 2013; pp. 644–654.
38. Peyre, P.; Aubry, P.; Fabbro, R.; Neveu, R.; Longuet, A. Analytical and numerical modelling of the direct metal deposition laser process. *J. Phys. D Appl. Phys.* **2008**, *41*, 025403. [CrossRef]

Journal of
Manufacturing and
Materials Processing

Article

Optimization of Laser Powder Bed Fusion Processing Using a Combination of Melt Pool Modeling and Design of Experiment Approaches: Density Control

Morgan Letenneur, Alena Kreitcberg and Vladimir Brailovski *

Department of Mechanical Engineering, École de technologie supérieure, 1100 Notre-Dame Street West, Montreal, QC H3C 1K3, Canada; morgan.letenneur.1@etsmtl.net (M.L.); alena.kreitcberg.1@ens.etsmtl.ca (A.K.)
* Correspondence: vladimir.brailovski@etsmtl.ca; Tel.: +1-514-396-8594

Received: 18 December 2018; Accepted: 12 February 2019; Published: 21 February 2019

Abstract: A simplified analytical model of the laser powder bed fusion (LPBF) process was used to develop a novel density prediction approach that can be adapted for any given powder feedstock and LPBF system. First, calibration coupons were built using IN625, Ti64 and Fe powders and a specific LPBF system. These coupons were manufactured using the predetermined ranges of laser power, scanning speed, hatching space, and layer thickness, and their densities were measured using conventional material characterization techniques. Next, a simplified melt pool model was used to calculate the melt pool dimensions for the selected sets of printing parameters. Both sets of data were then combined to predict the density of printed parts. This approach was additionally validated using the literature data on AlSi10Mg and 316L alloys, thus demonstrating that it can reliably be used to optimize the laser powder bed metal fusion process.

Keywords: additive manufacturing; laser powder bed fusion; process optimization; analytical model

---

1. Introduction

Interest in laser powder bed fusion (LPBF) additive manufacturing (AM) has spiked in many industries, creating a high demand for new AM-ready metallic materials [1]. However, the mechanical properties, surface finish, and precision of LPBF parts are dependent on more than 60 processing parameters [2], which all need to be optimized. There are currently two main ways to realize this process optimization for new alloys. Most often, this optimization is carried out by defining an experiment plan that covers different arrangements of laser power, scanning speed, hatching space, layer thickness, scanning strategy and part orientation for a given alloy [3–10]. Once the specimens are printed, their mechanical properties are evaluated and a conclusion is drawn on the influence of the different processing parameters on the final part geometric and service attributes. This approach yields satisfying results, but requires multiple printing jobs and time-consuming post-processing experiments. It could easily be realized for a single alloy, but becomes prohibitively expensive if multiple process optimization campaigns are required.

Another way a new AM material can be introduced is by applying a numerical modeling approach with the objective of finding the appropriate printing parameters, as shown in [11–15]. However, due to a large number of variables, these models require significant time and computer resources to model a single laser track, let alone a complex part. Moreover, the more complex the model, the more laborious the calibration procedure, which makes the process optimization more cumbersome and labor-intensive.

In this work, we investigate the possibility of using a combination of a simplified analytical model of the melt pool and of an experimental calibration routine to create a density control algorithm for the laser powder bed fusion process. The main objective of this approach is to reduce the time, the number of printing jobs and the quantity of post-processing characterization work needed to optimize the process for any given powder feedstock and any given LPBF system.

2. Methodology

Previous studies have demonstrated that the density of LPBF manufactured parts is mostly dependent on the following three dimensionless melt pool metrics (Figure 1): melt pool depth-to-layer thickness ratio (*D/t*), melt pool width-to-hatching space ratio (*W/h*), and melt pool length-to-melt pool width ratio (*L/W*), and that the highest density is generally obtained for 1.5 < *D/t* < 2, 1.5 < *W/h* < 2.5 and *L/W* < 2Pi [16,17]. Based on these observations, we investigated the possibility of using the *D/t*, *W/h* and *L/W* ratios to correlate a specific combination of LPBF processing parameters (laser power, scanning speed, layer thickness, and hatching space) with the density of a printed material.

This study was conducted in three phases: first, the analytical model of a thermal field generated by a moving heat source in a solid body is used to evaluate the melt pool dimensions for a given set of LPBF processing parameters. Then, a relationship between the melt pool dimensions and the density of the printed material was found experimentally for a given material and LPBF system. Finally, using the numerical model developed and the experimental relationship found, the LPBF processing parameters were linked to the density of the manufactured parts with the objective of developing a porosity prediction algorithm for different materials and different LPBF systems.

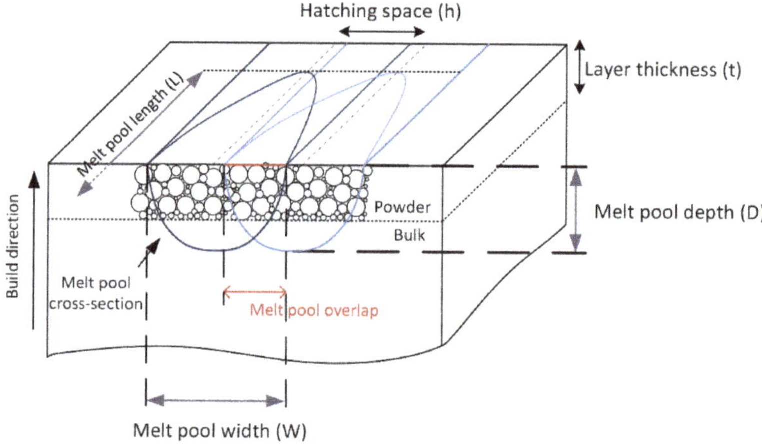

Figure 1. Schematic representation of the melt pool and the corresponding geometric characteristics [18].

2.1. Melt Pool Calculations

First, calculations of the LPBF melt pool dimensions were carried out using the analytical model of a semi-infinite solid with a moving Gaussian heat source [19]. This model has been successfully used for the determination of the LPBF processing parameters for pure iron [18] and Ti-Zr-Nb alloy [20]. The Gaussian model involves a symmetrical distribution of laser irradiance across the beam. The energy from the laser is assumed to be applied on the powder bed surface for a time interval defined by the scanning speed and the laser spot size. In this case, for a Gaussian beam moving with a given velocity, the temperature distribution $T(x{\cdot}y{\cdot}z)$ in the powder bed is calculated by Equations (1)–(3):

$$T(x{\cdot}y{\cdot}z) = T_0 + \frac{AP}{kr_f\pi^{\frac{3}{2}}} \int_{\infty}^{0} \frac{1}{1+\tau^2} \exp(C)d\tau \tag{1}$$

$$C = -\frac{\tau^2}{1+\tau^2}\left[\left(\xi - \frac{Pe}{2\tau^2}\right)^2 + \eta^2\right] - \tau^2\zeta^2 \tag{2}$$

$$\xi = \sqrt{2}\frac{x}{r_f} \cdot \eta = \sqrt{2}\frac{y}{r_f} \cdot \zeta = \sqrt{2}\frac{z}{r_f} \cdot Pe = \frac{r_f v}{2\sqrt{2}\alpha} \cdot \tau = \frac{r_f}{2\sqrt{2\alpha t}} \tag{3}$$

where T_0 is the powder bed temperature (°C); A, the absorptivity; P, the laser power (W); k, the thermal conductivity (W/(m·K)); r_f, the laser beam radius (m); Pe, the Peclet number; v, the scanning speed (m/s); α, the thermal diffusivity (m^2/s); ρ, the material density (kg/m^3); c_p, the specific heat (J/(kg·K)), and t, time (s).

The laser energy absorptivity A is estimated using Equation (4) from the Drude's theory [19,21]:

$$A \approx 0.365(\lambda\sigma_0)^{-0.5} = 0.365\left(\frac{\rho_0}{\lambda}\right)^{0.5} \tag{4}$$

where λ is the laser wavelength (μm), σ_0, the electrical conductivity (S/m), and ρ_0, the electrical resistivity of the irradiated material (Ohm·m).

2.2. Experimental Calibration

2.2.1. Materials, Equipment and Plan of Experiment

The experimental part of this study was conducted using the EOS-supplied IN625 powder and an EOSINT M280 LPBF system (EOS GmbH, Munich, Germany) equipped with a 400 W ytterbium fiber laser (beam radius r_f = 50 μm). The initial temperature of the substrate (build platform) T_0 = 60 °C. To design the plan of experiments, the analytical model represented by Equations (1)–(4) was used first. To this end, the following physical properties of an irradiated body: thermal conductivity k_0 [22], specific heat C_{p0} [23], and electrical resistivity ρ_0 [24] need to be calculated, taking into account the effective powder bed density φ, the latter being the powder morphology and spreading mechanism-dependent:

$$k = k_0 \times \frac{\varphi}{0.5(3-\varphi)}; \ C_p = C_{p0} \times \varphi; \ \rho = 0.696 \times \frac{4}{\varphi} \times \rho_0 \tag{5}$$

In this work, the density ϕ of IN625 powder spread by a standard EOS metal doctor blade and measured using the encapsulated samples method [25] was found to be close to 60%. Given the preceding, the IN625 alloy properties used for calculations are shown in Table 1. They were taken at room temperature, and it is considered that the preceding layer cools down to 60 °C between two scanning runs.

Table 1. Physical properties of IN625 powder [26,27].

	Bulk	Powder (φ = 60%)
Melting temperature, °C	1350	1350
Density, kg/m^3	8440	5072
Thermal conductivity, W/(m·K)	25.2	15.1
Specific heat capacity, J/(kg·K)	670	403
Electrical resistivity, 10^{-8} Ω·m	134	223

The temperature distribution map shown in Figure 2 represents an example of calculations using Equations (1)–(5). It corresponds to the following set of LPBF processing parameters: P = 270 W, v = 1000 mm/s, and t = 40 μm applied to IN625 powder (Table 1). From this temperature map, the melt pool width, depth and length are delimited by the alloy melting temperature of 1350 °C: W = 173 μm, D = 89 μm, and L = 806 μm, which correspond to the following dimensionless metrics: D/t = 2.2 and

L/W = 4.7 (since only a single track is modeled, the hatching space *h* is not considered at this stage). This calculation procedure can be repeated for any material and any given set of processing parameters.

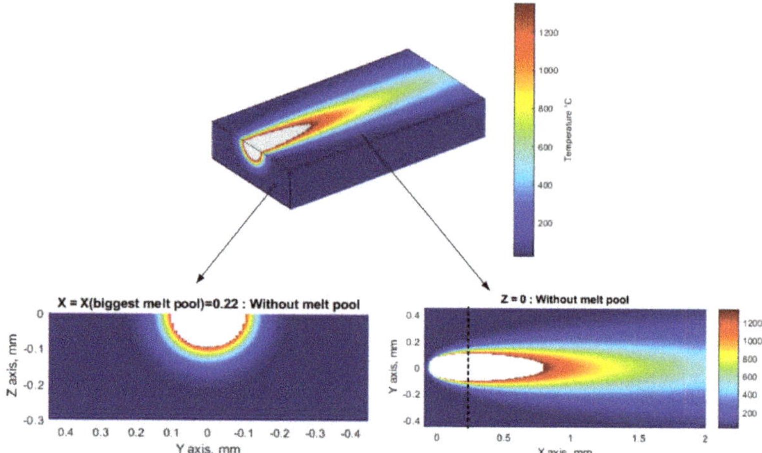

Figure 2. Melt pool dimensions for IN625 powder when P = 270 W, v = 1000 mm/s, and t = 40 μm; the melt pool width and depth are delimited by the alloy melting temperature.

2.2.2. Melt Pool Dimensions-Density Relationship

To establish the relationship between the previously defined dimensionless melt pool metrics and the density of manufactured parts, 10 mm-diameter 15 mm-height cylindrical coupons of IN625 alloy were printed to cover a *D/t* ratio ranging from 1 to 3.5, *W/h*, from 0.5 to 3 and *L/W*, from 3 to 6. To find the LPBF parameters resulting in these melt pool dimensions, the following ranges of printing parameters were reversely computed using the melt pool model presented in Section 2.1: the laser power varying from 160 to 350 W, the scanning speed, from 560 to 2800 mm/s, and the hatching space, from 30 to 550 μm; the layer thickness t was kept constant at 40 μm (see Table 2 for the selected values of the LPBF processing parameters). Two specimens were printed for each set of printing parameters.

Table 2. Imposed melt pool metrics and calculated processing parameters (plan of experiments).

		Melt Pool Dimensionless Metrics
	D/t	1, 1.5, 2, 2.5, 3, 3.5
Imposed	*W/h*	0.5, 1, 1.5, 2, 2.5, 3
	L/W	3.2, 3.9, 4.5, 5
		LPBF Processing Parameters
	Laser power *P*, W	160, 225, 340, 345, 350
Calculated	Scanning speed *v*, mm/s	560, 800, 1060, 1180, 1680, 1940, 2800
	Layer thickness *t*, μm	40
	Hatching space *h*, μm	30, 40, . . . , 180, 190, 210, 230, 270, 280, 350, 390, 430, 470, 550

After processing, the printed coupons were cut off the build plate and their densities measured using the Archimedes' technique (ASTM B962-15). Each density measurement using a SARTORIUS Secura 324-1s scale (Sartorius, Goettingen, Germany), having a precision of ~0.001 g, was repeated at least 3 times.

The results of this experiment are collected in Table 3 (Appendix A) and plotted in Figure 3 in the *D/t-W/h*-density coordinates. It can be seen from this figure that the density of IN625 coupons exceeding 99.5% (this value was selected arbitrarily to limit the amount of experimental data, while leaving enough space for optimization) was obtained for a *D/t* ratio ranging from 1.5 to 2.75 and a *W/h* ratio ranging from 1.8 to 2.8. The corresponding *L/W* ratio ranged from 3.8 to 4.6 (not shown on this diagram). Note that the calculated values fall close to the ranges recommended in the literature, which are 1.5 < *D/t* < 2, 1.5 < *W/h* < 2.5 and *L/W* < 2Pi [28].

From Figure 3, assuming that the calculated *D/t*, *W/h* ratios correspond to the effectively obtained melt pool dimensions, the materials density can be expressed as their function as follows:

$$\rho = a0 + a1 \cdot \left(\frac{D}{t}\right) + a2 \cdot \left(\frac{W}{h}\right) + a3 \cdot \left(\frac{D}{t}\right)^2 + a4 \cdot \left(\frac{D}{t}\right) \cdot \left(\frac{W}{h}\right) + a5 \cdot \left(\frac{W}{h}\right)^2 \tag{6}$$

where $a0 = 0.512$, $a1 = 0.212$, $a2 = 0.225$, $a3 = -0.027$, $a4 = -0.384$, and $a5 = -0.031$.

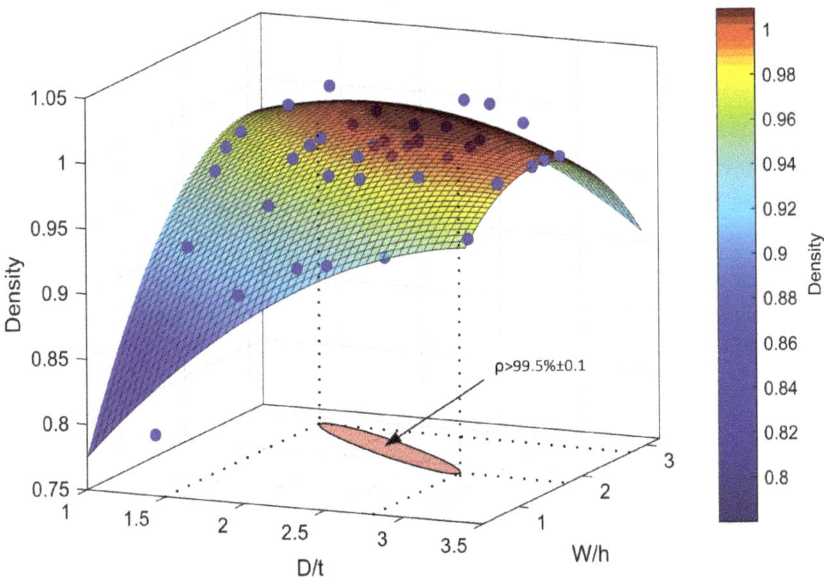

Figure 3. Density of the printed coupons as a function of the *D/t* and *W/h* ratios; the calculated *D/t-W/h* area corresponds to the measured density of the printed material exceeding 99.5 ± 0.1%.

2.3. Energy Density-Build Rate Processing Map

In this work, the LPBF processing conditions were expressed by a combination of two metrics: the volumetric laser energy density E (J/mm^3) (7) and the material build rate BR (cm^3/h) (8); the product of both corresponds to the laser power P (Watts).

$$E\left(\text{J/mm}^3\right) = \frac{P}{v \cdot h \cdot t} \tag{7}$$

$$BR\left(cm^3/h\right) = v \cdot h \cdot t \tag{8}$$

Next, the analytical model (1–5) and Table 1 were used to map three *E–BR* areas corresponding to the experimentally obtained optimal ranges of the melt pool metrics (Figure 4a): *D/t* = 1.5–2.75, *W/h* = 1.8–2.8, and *L/W* = 3.8–4.6. These maps are calculated by varying the laser power from 20 to 380 W; the scanning speed, from 100 to 4000 mm/s; the hatching space, from 30 to 200 μm, and the layer thickness, from 20 to 80 μm.

Three *E–BR* areas of Figure 4a were then superposed in Figure 4b to schematically delimit a common processing window, which must guarantee the maximum density of printed IN625 parts. Next, the densities of the printed coupons (Figure 3 and Table 1 in the Annex) were superposed on this processing window, and it can be seen that the coupons with a density $\geq$99.5% are indeed located, within a certain margin of error, in the numerically predicted optimal processing window. By refining the scanning steps and using calibration Equation (6), the described approach can then be used to build a more detailed processing map for IN625 powder (Figure 4c).

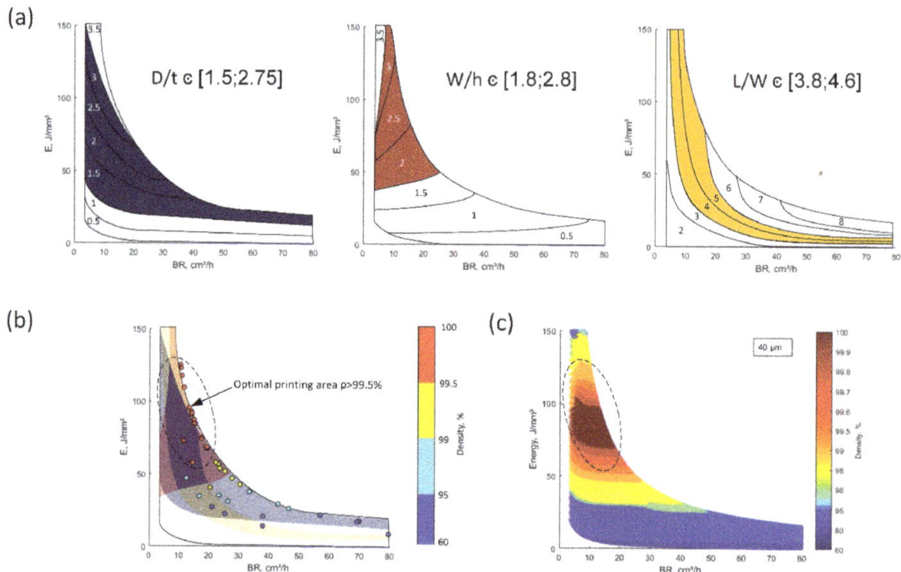

Figure 4. (**a**) Optimal areas for the *D/t*, *W/h*, and *L/W* ratios; (**b**) superposition of the numerically optimized processing window; (**c**) experimentally calibrated processing map.

2.4. Validation Strategy for the Proposed Processing Optimization Approach

We hypothesize that such a combined processing optimization approach is valid for any material processed by a given LPBF system. In our case it is the EOS M280 LPBF system. To verify this hypothesis, the results of such an optimization were compared with the numerical and experimental data found in the literature. This comparison was carried out in two phases: first, the melt pool dimensions were calculated for AlSi10Mg and 316L powders using a simplified analytical model of this work (Equations (1)–(5)), and then compared against those obtained for the same feedstock material, but using more powerful finite element models (FEM). These FEM models take into account the optical penetration depth, the mass transfer-related phenomena, such as the Marangoni convection and the Rayleigh capillary flow, and the heat losses to the environment [11,29]. Secondly, the numerically predicted densities for pure iron and Ti-6Al-4V alloy powders are compared with their experimentally measured equivalents from [6] and [18]. The optimal processing windows for all these validation studies were obtained using the algorithm previously presented. The data used for these calculations were taken from the corresponding literature sources.

3. Validation

3.1. Melt Pool Dimensions (Single Track)

As experimental validation of the melt pool model used in this study (Equations (1)–(5)) has already been carried out in our previous work [20], a decision was made to extend the validation experiments to literature data. With this objective in mind, the single track melt pool dimensions calculated by the said analytical model were compared with those calculated by two different finite element models. Note that each of these models was experimentally validated by their authors for two different alloys: 316L [11] and AlSi10Mg [14]. For ease of understanding and because the width and the depth of the melt pool are the most important characteristics impacting the density of the printed material, the geometric validation was limited to these two characteristics.

Regarding 316L, the simulations were carried out with an initial powder bed temperature of 296 K (first track in [11]), a fixed laser power of 110 W and a scanning speed ranging from 80 to 150 mm/s. For AlSi10Mg, the simulations were realized for a laser power ranging from 150 to 300 W and a fixed scanning speed of 200 mm/s [14]. In both cases, computations using (Equations (1)–(5)) were carried out using the physical properties taken from the corresponding literature sources (Table 3). In other words, the electrical resistivity, the thermal conductivity and the specific heat capacity values used in our calculations were set identical to those used in [11] and [14] and recalculated for a 60% powder bed density. This last value was selected on the basis of our previous results because no such information was provided in the literature sources.

Table 3. Physical properties of the AlSi10Mg [13] and 316L [15] powders used for melt pool modeling.

	AlSi10Mg		316L	
	Bulk	**Powder (φ = 60%)**	**Bulk**	**Powder (φ = 60%)**
Melting temperature, °C	600	600	1400	1400
Density, kg/m^3	2650	1590	8000	4800
Thermal conductivity, W/(m·K)	147	88.2	16.2	9.7
Specific heat capacity, J/(kg·K)	739	443.4	530	318
Electrical resistivity, 10^{-8} Ω·m	7.8 [30]	4.7	74	123

The melt pool profiles calculated by the model of this study and those found in the literature are plotted in Figure 5 for different sets of printing parameters. The mean deviations between the results of the analytical and the finite element models are 4.3 ± 1.2% for 316L and 10.5 ± 5.8% for AlSi10Mg. (Note that the numerically estimated impact of a 5.8% deviation in the AlSi10Mg melt pool dimensions would have introduced only ~0.2% variation in the predicted density values; the last number being calculated by introducing a value of 5.8% in the density Equation (6)).

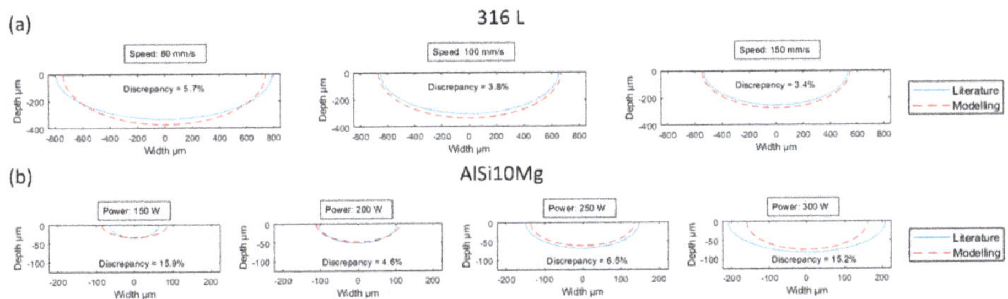

Figure 5. Comparison of the melt pools profiles for (**a**) 316L [11] and (**b**) AlSi10Mg [14] alloys.

3.2. Density of the Same Alloy Printed with Two Different Layer Thicknesses

Using a combination of the analytical modeling of melt pool dimensions (Equations (1)–(5)) and Equation (6), representing the correspondence between the melt pool dimensions and the density of the printed parts, processing windows can be calculated for multi-track LPBF. To validate this approach, processing maps for Ti64 alloy printed with two layer thicknesses (30 and 60 um) are plotted in Figure 6. The physical properties used for these calculations are collected in Table 4.

Note that the powder bed densities (evaluated using the method of encapsulated samples [25]) in these two cases are not identical: it is higher in the former than in the latter case, with φ = 70% for t = 30 µm and φ = 60% for t = 60 µm. These changes in the powder bed density are due to the differences in the powder spreading conditions for different layer thicknesses as demonstrated in [31]. If a layer thickness is smaller than the D90 value of the powder particle distribution [32], the density increases because the biggest particles are kept at the top of the layer and finally swept out by the recoater [33], which increases the powder bed density. It can be seen from Figure 6 that the higher the powder bed density (70%, Figure 6b, instead of 60%, Figure 6a), the higher the optimal laser power density, while the lower the build rate of the process. Similar results were reported in [31].

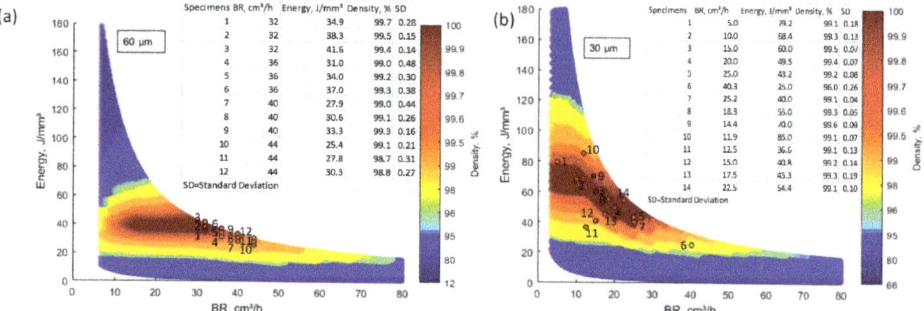

Figure 6. Processing maps for Ti64 alloy for layer thicknesses of (**a**) 60μm and (**b**) 30μm (EOS M 280).

Table 4. Physical properties of Ti64 powders used for melt pool modeling [26,34].

	Ti64 (t = 60 μm)		Ti64 (t = 30 μm)
	Bulk	**Powder (φ = 60%)**	**Powder (φ = 70%)**
Melting temperature, °C	1660	1660	1660
Density, kg/m^3	4410	2630	3150
Thermal conductivity, W/(m·K)	7.3	4.27	5.18
Specific heat capacity, J/(kg·K)	570	342	405
Electrical resistivity, 10^{-8} Ω·m	170	283	239

The experimentally measured densities of Ti64 coupons printed with the layer thicknesses of t = 60 μm (Figure 6a) and 30 μm (Figure 6b) were then superposed on the calculated processing maps (these coupons were printed using EOS Ti64 powder and the EOS M280 system of this study). The mean porosity deviations for t = 60 μm corresponded to 0.8%, while for t = 30 μm, it was 0.4%.

3.3. Density of Two Different Alloys

The reliability of the proposed processing optimization approach was then studied for Fe [18] and AlSi10Mg [35] powders. The density predictions were made using the physical properties of Table 5 and the processing maps are plotted in Figure 7a,b. The experimentally measured density values were superposed, and the deviations between the model and the experiment corresponded to 0.8% for Fe and 0.7% for AlSi10Mg powders.

Table 5. Physical properties of Fe [3] and AlSi10Mg [20] powders used for melt pool modeling.

	Fe		AlSi10Mg	
	Bulk	**Powder (t = 60 μm; φ = 60%)**	**Bulk**	**Powder (t = 30 μm; φ = 45%)**
Melting temperature, °C	1660	1660	600	600
Density, kg/m^3	8000	4800	2650	1200
Thermal conductivity, W/(m·K)	16.2	9.7	147	66.6
Specific heat capacity, J/(kg·K)	530	318	739	335
Electrical resistivity, 10^{-8} Ω·m	74	123	7.8	17.2

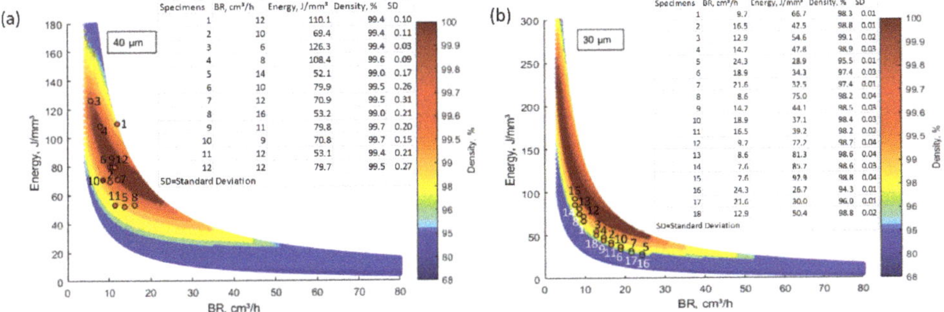

Figure 7. Processing maps for (**a**) Fe and (**b**) AlSi10Mg.

4. Discussion and Application Example

Notwithstanding that the simplified analytical model used in this study does not take into account the specificities of a given printing system in terms of heat exchange and powder spreading conditions, which both influence the density of the manufactured parts, it was demonstrated that such a model could provide useful information in terms of the energy density and the build rate values, which are potentially suitable for the printing of dense parts. However, to determine the exact set of processing parameters, such as the laser power, speed, hatching space, and layer thickness, an additional condition must be respected, and this condition corresponds to the ratio between the hatching space and the layer thickness, h/t.

To establish such a condition, the previously developed model was used to plot the density of IN625 components as a function of the h/t ratio for different layer thicknesses (Figure 8a). From this plot, it is clear that to maximize the material density, the selection of a hatching space must be related to the selection of a layer thickness (Figure 8b). For example, to guarantee the maximum material density $\geq$99.8%, with a layer thickness of t = 30 μm, the hatching space variations must be limited to the 50 to 80 μm range, while for a layer thickness of t = 90 μm, the hatching space variations must be limited to the 110 to 220 μm range.

These results are shown in Figure 8b to present the h-t area corresponding to the maximum density of IN625 parts. In the same figure, corresponding h-t areas are plotted for AlSi10Mg (Table 3) and Ti64 (Table 4) alloys, for comparison. Finally, once obtained, such plots provide guidance for the selection of the most appropriate hatching space/layer thickness combinations.

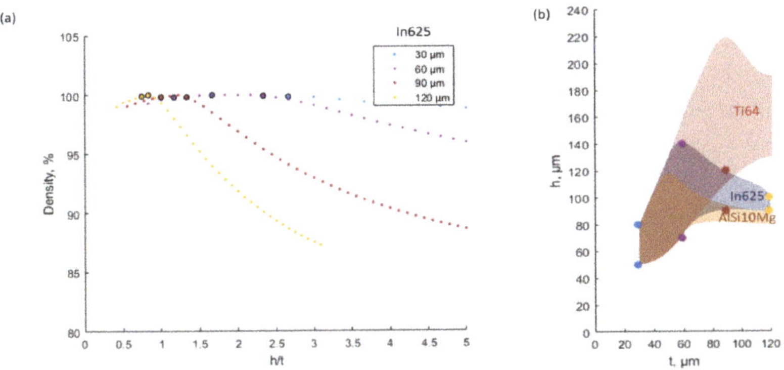

Figure 8. Hatching space/Layer thickness relations: (**a**) density as a function of the h/t ratio; (**b**) hatching space as a function of the layer thickness for maximum density $\geq$99.5% (IN625, AlSi10Mg and Ti64 powders for an EOS M280 LPBF system).

Note that even though this combined modeling-experiment approach was validated for only one specific LPBF system (EOS M 280), we hypothesize that it can be extended to any LPBF system, provided an adequate calibration experiment is carried out. To this end, after generating the first processing map assuming the physical properties of the material taken from the literature and a powder bed density of 60%, a series of calibration coupons must be printed. Once the density of the printed coupons is measured, the model must be adjusted to fit the experimentally obtained values, by modifying the coefficients of Equation 6. Finally, the relation between the hatching space and the layer thickness can be plotted for a maximum printed material density (see Figure 8b). Once all these conditions are met, the LPBF processing parameters (laser powder, scanning speed, hatching space and layer thickness) can be determined using the following protocol:

1. The layer thickness is selected first to provide a required precision/performance relationship (Figure 9a). If we take $t = 40$ μm to favor precision, the processing map for this layer thickness can be built as shown in Figure 9c.
2. As, in this case, h can vary from 50 to 110 μm (Figure 9b), the highest hatch value of 110 μm can be specified to improve the process productivity.
3. To print components with a material density $\geq$99.8% and a maximum allowable build rate of $BR = 15$ cm^3/h (Figure 9), the corresponding volumetric laser energy density corresponds to $E = 67$ J/mm^3 (see the dot in Figure 9c). Since $t = 40$ μm and $h = 110$ μm, the remaining LPBF parameters can easily be defined using the energy density and build rate definitions of Equations (7) and (8): laser power ~285 W and scanning speed ~960 mm/s [36].

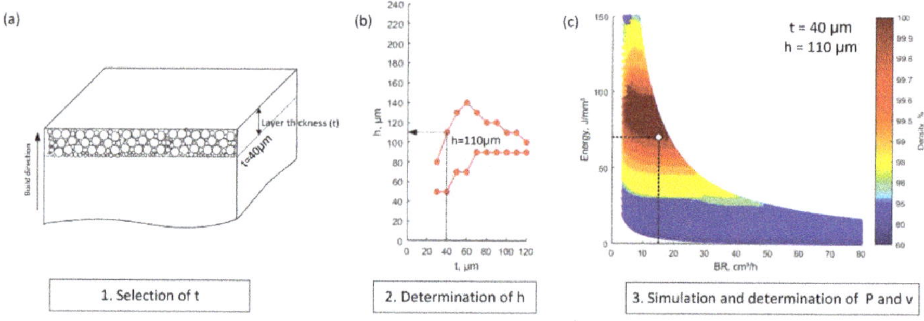

Figure 9. Steps needed for the printing parameters determination from a processing map: (**a**) selection of a layer thickness, (**b**) selection of an appropriate hatching space, and (**c**) determination of the corresponding laser power and scanning speed values.

As it is widely assumed that the smaller the layer thickness, the better the surface finish and part precision, but the lower the build rate, it is recommended to work with layer thicknesses of 30 or 40 μm when precision is required, and of 50 or 60 μm, when process productivity is more important.

5. Conclusions

A simplified analytical model of the LPBF process was used to develop the density prediction algorithm for a given powder feedstock and a given LPBF system. Using a set of density calibration coupons built with the laser power varying from 160 to 350 W; the scanning speed, from 500 to 2800 mm/s; the hatching space, from 30 to 550 μm, and the layer thickness, from 30 to 60 μm, this model was adapted for the IN625 alloy powder and an M280 EOS LPBF system. This approach was then validated for different alloys and processing conditions using literature data, thus demonstrating its potential for the LPBF process optimization. It was also shown that the layer thickness value has a direct influence on the creation of the processing map and must be taken into account during the calibration step. This step represents a must-follow requirement to improve the prediction capability of the model because it takes into account the specificities of a given LPBF system related to particular powder recoating and heat transfer conditions, which differ from one printer to another, and influence the final density of the manufactured parts. However, once calibrated for a selected LPBF system, the model could be used for different alloys processed with this same system, avoiding trial-and-error-based process optimization routines.

To further this work, a comprehensive study should be conducted to evaluate the influence of parts' geometry, size, orientation and support on the density predictions.

Author Contributions: The work plan was developed by M.L. to meet the study objectives defined by himself and V.B. The specimen design and the alloy selection were carried out by all the coauthors based upon the state of the art, previous research of V.B., and the current trends in the field. The design, fabrication and testing of the specimens were performed by M.L., V.B. contributed to the data organization, results interpretation and manuscript redaction.

Funding: This research received no external funding.

Acknowledgments: The authors would like to express their appreciation for the financial support provided by NSERC (Natural Sciences and Engineering Research Council of Canada) and for the technical support provided by M. Samoilenko.

Conflicts of Interest: The authors declare no conflict of interest.

Appendix A

Table A1. Printing parameters, melt pool geometry and density of IN625 test coupons (M280 EOS).

Spec.	Power, W	Speed, m/s	Hatching space, mm	Energy density, J/mm³	Build rate, cm³/h	Mel pool geometry			Density, %	SD ±%
						D/t	*W/h*	*L/W*		
1	160	2.8	0.092	15.6	37.0	1	1	5.7	78.3	0.7
2	160	2.8	0.061	23.3	24.7	1	1.5	5.7	87.6	0.2
3	160	2.8	0.052	27.4	21.0	1	1.76	5.7	91.6	0.3
4	160	2.8	0.046	31.1	18.5	1	2	5.7	95.6	0.1
5	160	2.8	0.037	38.9	14.8	1	2.5	5.7	95.6	0.1
6	160	2.8	0.031	46.7	12.3	1	3	5.7	97.7	0.1
7	225	1.94	0.265	10.9	74.1	1.5	0.5	5.7	79.03	1.1
8	225	1.94	0.133	21.9	37.1	1.5	1	5.7	92.5	0.2
9	225	1.94	0.088	32.8	24.7	1.5	1.5	5.7	97.3	0.1
10	225	1.94	0.075	38.5	21.1	1.5	1.76	5.7	98.7	0.1
11	225	1.94	0.066	43.7	18.5	1.5	2	5.7	99.5	0.1
12	225	1.94	0.053	54.6	14.8	1.5	2.5	5.7	99.6	<0.1
13	225	1.94	0.044	65.6	12.4	1.5	3	5.7	99.6	<0.1
14	350	1.68	0.347	15.0	83.9	2	0.5	5.4	85.0	0.3
15	350	1.68	0.173	30.0	42.0	2	1	5.4	96.4	0.1
16	350	1.68	0.116	45.0	28.0	2	1.5	5.4	99.1	<0.1
17	350	1.68	0.099	52.8	23.8	2	1.76	5.4	99.4	<0.1
18	350	1.68	0.087	60.0	21.0	2	2	5.4	99.6	<0.1
19	350	1.68	0.069	75.1	16.8	2	2.5	5.4	99.4	0.1
20	350	1.68	0.058	90.1	14.0	2	3	5.4	99.6	0.1
21	350	1.18	0.388	19.1	65.9	2.45	0.5	5.0	93.0	0.6
22	350	1.18	0.194	38.2	32.9	2.45	1	5.0	98.9	0.1
23	350	1.18	0.129	57.4	22.0	2.45	1.5	5.0	99.3	<0.1
24	350	1.18	0.110	67.3	18.7	2.45	1.76	5.0	99.5	0.1
25	350	1.18	0.097	76.5	16.5	2.45	2	5.0	99.4	0.1
26	350	1.18	0.078	95.6	13.2	2.45	2.5	5.0	99.6	0.1
27	350	1.18	0.065	114.7	11.0	2.45	3	5.0	99.7	0.2
28	345	1.06	0.429	19.0	65.4	2.5	0.5	4.5	93.3	0.1
29	345	1.06	0.214	38.0	32.7	2.5	1	4.5	98.7	0.2
30	345	1.06	0.143	57.0	21.8	2.5	1.5	4.5	99.3	0.2
31	345	1.06	0.122	66.8	18.6	2.5	1.76	4.5	99.4	0.1
32	345	1.06	0.107	75.9	16.4	2.5	2	4.5	99.4	<0.1
33	345	1.06	0.086	94.9	13.1	2.5	2.5	4.5	99.5	<0.1
34	345	1.06	0.071	113.9	10.9	2.5	3	4.5	99.7	0.1
35	350	0.8	0.469	23.3	54.1	3	0.5	3.9	94.3	0.1
36	350	0.8	0.235	46.6	27.0	3	1	3.9	99.1	0.1
37	350	0.8	0.156	69.9	18.0	3	1.5	3.9	99.5	<0.1
38	350	0.8	0.133	82.0	15.4	3	1.76	3.9	99.5	0.1

Table A1. *Cont.*

Spec.	Power, W	Speed, m/s	Hatching space, mm	Energy density, J/mm^3	Build rate, cm^3/h	Mel pool geometry			Density, %	SD ±%
						D/t	*W/h*	*L/W*		
39	350	0.8	0.117	93.2	13.5	3	2	3.9	99.7	0.1
40	350	0.8	0.094	116.5	10.8	3	2.5	3.9	99.4	0.1
41	340	0.56	0.551	27.5	44.4	3.5	0.5	3.4	96.3	0.1
42	340	0.56	0.276	55.1	22.2	3.5	1	3.4	99.4	0.1
43	340	0.56	0.184	82.6	14.8	3.5	1.5	3.4	99.4	0.2
44	340	0.56	0.157	97.0	12.6	3.5	1.76	3.4	99.5	0.1
45	340	0.56	0.138	110.2	11.1	3.5	2	3.4	99.2	0.2

References

1. Huang, Y.; Leu, M.C.; Mazumder, J.; Donmez, A. Additive manufacturing: Current state, future potential, gaps and needs, and recommendations. *J. Manuf. Sci. Eng.* **2015**, *137*, 014001. [CrossRef]

2. Brandt, M. *Laser Additive Manufacturing: Materials, Design, Technologies, and Applications*; Woodhead Publishing: Cambridge, UK, 2016.

3. Kasperovich, G.; Haubrich, J.; Gussone, J.; Requena, G. Correlation between porosity and processing parameters in TiAl6V4 produced by selective laser melting. *Mater. Des.* **2016**, *105*, 160–170. [CrossRef]

4. Delgado, J.; Ciurana, J.; Rodríguez, C.A. Influence of process parameters on part quality and mechanical properties for DMLS and SLM with iron-based materials. *Int. J. Adv. Manuf. Technol.* **2012**, *60*, 601–610. [CrossRef]

5. Kurzynowski, T.; Chlebus, E.; Kuźnicka, B.; Reiner, J. Parameters in Selective Laser Melting for processing metallic powders. In Proceedings of the High Power Laser Materials Processing: Lasers, Beam Delivery, Diagnostics, and Applications, San Francisco, CA, USA, 24–26 January 2012; Volume 8239.

6. Bartolomeu, F.; Faria, S.; Carvalho, O.; Pinto, E.; Alves, N.; Silva, F.S.; Miranda, G. Predictive models for physical and mechanical properties of Ti6Al4V produced by Selective Laser Melting. *Mater. Sci. Eng. A* **2016**, *663*, 181–192. [CrossRef]

7. Ahmed, A.; Wahab, M.S.; Raus, A.A.; Kamarudin, K.; Bakhsh, Q.; Ali, D. Effects of Selective Laser Melting Parameters on Relative Density of AlSi10Mg. *Int. J. Eng. Technol.* **2016**, *8*, 2552–2557. [CrossRef]

8. Cherry, J.A.; Davies, H.M.; Mehmood, S.; Lavery, N.P.; Brown, S.G.; Sienz, J. Investigation into the Effect of process parameters on microstructural and physical properties of 316L stainless steel parts by selective laser melting. *Int. J. Adv. Manuf. Technol.* **2014**, *76*, 869–879. [CrossRef]

9. Rashid, R.; Masood, S.H.; Ruan, D.; Palanisamy, S.; Rahman Rashid, R.A.; Elambasseril, J.; Brandt, M. Effect of energy per layer on the anisotropy of selective laser melted AlSi12 aluminium alloy. *Addit. Manuf.* **2018**, *22*, 426–439. [CrossRef]

10. Rashid, R.; Masood, S.H.; Ruan, D.; Palanisamy, S.; Rashid, R.R.; Brandt, M. Effect of scan strategy on density and metallurgical properties of 17-4PH parts printed by Selective Laser Melting (SLM). *J. Mater. Process. Technol.* **2017**, *249*, 502–511. [CrossRef]

11. Foroozmehr, A.; Badrossamay, M.; Foroozmehr, E. Finite element simulation of selective laser melting process considering optical penetration depth of laser in powder bed. *Mater. Des.* **2016**, *89*, 255–263. [CrossRef]

12. Fischer, P.; Karapatis, N.; Romano, V.; Glardon, R.; Weber, H.P. A model for the interaction of near-infrared laser pulses with metal powders in selective laser sintering. *Appl. Phys. A* **2002**, *74*, 467–474. [CrossRef]

13. Fischer, P.; Romano, V.; Weber, H.P.; Karapatis, N.P.; Boillat, E.; Glardon, R. Sintering of commercially pure titanium powder with a Nd: YAG laser source. *Acta Mater.* **2003**, *51*, 1651–1662. [CrossRef]

14. Li, Y.; Gu, D. Parametric analysis of thermal behavior during selective laser melting additive manufacturing of aluminum alloy powder. *Mater. Des.* **2014**, *63*, 856–867. [CrossRef]

15. Zhang, Z.; Huang, Y.; Kasinathan, A.R.; Shahabad, S.I.; Ali, U.; Mahmoodkhani, Y.; Toyserkani, E. 3-Dimensional heat transfer modeling for laser powder-bed fusion additive manufacturing with volumetric heat sources based on varied thermal conductivity and absorptivity. *Opt. Laser Technol.* **2019**, *109*, 297–312. [CrossRef]

16. Yadroitsev, I.; Yadroitsava, I.; Bertrand, P.; Smurov, I. Philippe Bertrand and Igor Smurov, Factor analysis of selective laser melting process parameters and geometrical characteristics of synthesized single tracks. *Rapid Prototyp. J.* **2012**, *18*, 201–208. [CrossRef]

17. Aboulkhair, N.T.; Everitt, N.M.; Ashcroft, I.; Tuck, C. Reducing porosity in AlSi10Mg parts processed by selective laser melting. *Addit. Manuf.* **2014**, *1*, 77–86. [CrossRef]
18. Letenneur, M.; Brailovski, V.; Kreitberg, A.; Paserin, V.; Bailon-Poujol, I. Laser powder bed fusion of water-atomized iron-based powders: Process optimization. *J. Manuf. Mater. Process.* **2017**, *1*, 23. [CrossRef]
19. Schuöcker, D. *Handbook of the Eurolaser Academy*; Springer Science & Business Media: Vienna, Austria, 1998; Volume 2.
20. Kreitberg, A.; Brailovski, V.; Prokoshkin, S. New biocompatible near-beta Ti-Zr-Nb alloy processed by laser powder bed fusion: Process optimization. *J. Mater. Process. Technol.* **2018**, *252*, 821–829. [CrossRef]
21. Hagen, E.; Rubens, H. Über Beziehungen des Reflexions-und Emissionsvermögens der Metalle zu ihrem elektrischen Leitvermögen. *Ann. Phys.* **1903**, *316*, 873–901. [CrossRef]
22. Bala, K.; Pradhan, P.R.; Saxena, N.S.; Saksena, M.P. Effective thermal conductivity of copper powders. *J. Phys. D Appl. Phys.* **1989**, *22*, 1068. [CrossRef]
23. Sumirat, I.; Ando, Y.; Shimamura, S. Theoretical consideration of the effect of porosity on thermal conductivity of porous materials. *J. Porous Mater.* **2006**, *13*, 439–443. [CrossRef]
24. Liu, P.; Fu, C.; Li, T. Calculation formula for apparent electrical resistivity of high porosity metal materials. *Sci. China Ser. E Technol. Sci.* **1999**, *42*, 294–301. [CrossRef]
25. Jacob, G.; Donmez, A.; Slotwinski, J.; Moylan, S. Measurement of powder bed density in powder bed fusion additive manufacturing processes. *Meas. Sci. Technol.* **2016**, *27*, 115601. [CrossRef]
26. American Society for Metals. *Metals Handbook. 2. Properties and Selection: Nonferrous Alloys and Special-Purpose Materials*; American Society for Metals: Materials Park, OH, USA, 1990.
27. Davis, J.R. *ASM Specialty Handbook: Heat-Resistant Materials*; ASM International: Materials Park, OH, USA, 1997.
28. Gunenthiram, V.; Peyre, P.; Schneider, M.; Dal, M.; Coste, F.; Fabbro, R. Analysis of laser-melt pool-powder bed interaction during the selective laser melting of a stainless steel. *J. Laser Appl.* **2017**, *29*, 022303. [CrossRef]
29. King, W.E.; Anderson, A.T.; Ferencz, R.M.; Hodge, N.E.; Kamath, C.; Khairallah, S.A.; Rubenchik, A.M. Laser powder bed fusion additive manufacturing of metals; physics, computational, and materials challenges. *Appl. Phys. Rev.* **2015**, *2*, 041304. [CrossRef]
30. Silbernagel, C.; Ashcroft, I.; Dickens, P.; Galea, M. Electrical resistivity of additively manufactured AlSi10Mg for use in electric motors. *Addit. Manuf.* **2018**, *21*, 395–403. [CrossRef]
31. Abd-Elghany, K.; Bourell, D. Property evaluation of 304L stainless steel fabricated by selective laser melting. *Rapid Prototyp. J.* **2012**, *18*, 420–428. [CrossRef]
32. Spierings, A.B.; Levy, G. Comparison of density of stainless steel 316L parts produced with selective laser melting using different powder grades. In Proceedings of the Annual International Solid Freeform Fabrication Symposium, Austin, TX, USA, 3–5 August 2009.
33. Jacob, G.; Brown, C.U.; Donmez, M.A. *The Influence of Spreading Metal Powders with Different Particle Size Distributions on the Powder Bed Density in Laser-Based Powder Bed Fusion Processes*; Advanced Manufacturing Series (NIST AMS) 100-17; National Institute of Standards and Technology (NIST): Gaithersburg, MD, USA, 2018.
34. Polmear, I.J. *Light Alloys: From Traditional Alloys to Nanocrystals*, 4th ed.; Elsevier: Oxford, UK, 2005.
35. Krishnan, M.; Atzeni, E.; Canali, R.; Calignano, F.; Manfredi, D.; Ambrosio, E.P.; Iuliano, L. On the effect of process parameters on properties of AlSi10Mg parts produced by DMLS. *Rapid Prototyp. J.* **2014**, *20*, 449–458. [CrossRef]
36. Poulin, J.-R.; Letenneur, M.; Terriault, P.; Brailovski, V. Influence of intentionally-induced porosity and post-processing conditions on the mechanical properties of laser powder bed fused inconel 625. In *Selected Technical Papers (STP1620), Proceedings of ASTM Symposium on Structural Integrity of Additive Manufactured Parts, Washington, DC, USA, 6–8 November 2018*; ASTM International: West Conshohocken, PA, USA, 2018.

Journal of
*Manufacturing and
Materials Processing*

Article

Discrete Element Simulation of Orthogonal Machining of Soda-Lime Glass with Seed Cracks

Guang Yang, Hazem Alkotami and Shuting Lei *

Department of Industrial and Manufacturing Systems Engineering, Kansas State University, Manhattan,
KS 66506, USA; gyang@ksu.edu (G.Y.); alkotami@ksu.edu (H.A.)
* Correspondence: lei@ksu.edu; Tel.: +1-785-532-3731

Received: 30 December 2019; Accepted: 13 January 2020; Published: 16 January 2020

Abstract: Demands for producing high quality glass components have been increasing due to their superior mechanical and optical properties. However, due to their high hardness and brittleness, they present great challenges to researchers when developing new machining processes. In this work, the discrete element method (DEM) is used to simulate orthogonal machining of synthetic soda-lime glass workpieces that are created using a bonded particle model and installed with four different types of seed cracks. The effects of these seed cracks on machining performance are studied and predicted through the DEM simulation. It is found that cutting force, random cracks, and surface roughness are reduced by up to 90%, 74%, and 47%, respectively, for the workpieces with seed cracks compared to the regular ones. The results show that high performance machining through DEM simulation can be achieved with optimal seed cracks.

Keywords: discrete element method; orthogonal cutting; seed cracks; surface roughness

---

1. Introduction

Glass materials have been widely used in our daily life due to their superior properties, but machining these materials has always been a challenge due to their hard and brittle nature. There are various machining techniques such as turning, milling, drilling, grinding, and laser machining. However, it is very difficult to study the complex process of crack initiation and propagation through experimental observation and theoretical analysis. In order to understand the dynamics of random crack initiation and propagation in the glass cutting process, discrete element method is adopted to model and simulate the cutting process [1].

The discrete element method (DEM) is a numerical technique which models solid structures as bonded particles. These particles can deform and displace from one another and interact through contacts or interfaces between them. Unlike the finite element method (FEM), the discrete element method has the advantage of modeling brittle fracture. It can describe nonlinear behavior of brittle materials and handle the complex particle contact physical process with coupled shear and bulk deformation effects. Since the DEM was first introduced by Cundall [2], it has been widely applied in various areas such as simulating crushable soil [3], granular flow [4], and even behavior of the earthquakes [5]. In recent years, the DEM also has been used in simulating the cutting process of various materials such as rock, ceramics, and carbon fiber reinforced polymer [6–8].

Nowadays, hybrid machining has become more and more popular compared to traditional machining methods. It combines different machining actions on the material that need to be removed and makes use of the combined advantages to avoid or reduce some adverse effects [9]. For example, vibration-assisted machining combines machining with small-amplitude tool vibrations. During this process, the cutting tool loses contact with the chips on a specified amplitude, resulting in decreased machining forces and improved tool life and surface finish [10]. Chemical-assisted micromachining combines micromachining and chemical reaction of the workpiece. For silicon, the bonding forces between Si particles on the surfaces can be weakened by the hydrofluoric acid. A low concentration of hydrofluoric acid added to the abrasive slurry in ultrasonic machining can increase the material removal rate and surface quality [11]. Laser-induced crack-assisted machining combines laser micromachining and traditional orthogonal cutting. Before orthogonal cutting is carried out, the workpiece is treated by a femtosecond laser to induce seed cracks on the workpiece. This process is able to reduce cutting force, subsurface damage, and tool wear [12]. Due to the hard and brittle nature, it is very difficult to study the complex process of crack initiation and propagation through experimental observations and theoretical analysis. It also brings big challenges to researchers when developing new machining processes. Therefore, the DEM simulation is used in this study to help us understand the dynamics of random crack initiation and propagation during the cutting process.

In this paper, the discrete element method is used to simulate the orthogonal cutting of soda-lime glass with different seed cracks. The purpose is to predict the effects of different seed cracks on the cutting process. The first step is to create a synthetic material that behaves like soda-lime glass. Then, the macro-properties are calibrated by adjusting the micro-parameters of the DEM model to match the mechanical properties of the real soda-lime glass. Orthogonal cutting experiments are conducted in order to validate the cutting forces, subsurface cracks, and chips. Finally, the cutting simulations with four different types of seed cracks are conducted in order to optimize the cutting force, random cracks, and surface roughness. Through this study, the effects of different types of seed cracks during the orthogonal cutting is predicted.

2. Model Creation and Validation

Particle flow code in two dimensions (PFC2D), a program based on DEM, is used to model soda-lime glass and random distributions of circular particles are adopted to achieve an isotropic material. The particle contact behaviors of the model are expressed by flat-joint bonds, which is detailed in the authors' previous work [1]. The interactions and movements of the circular particles can be simulated by PFC2D. It allows finite displacement and rotations of discrete bodies, including complete detachment, and recognizes new contacts automatically as the calculation progresses.

2.1. Creation of the Specimen of Bonded Particles

In order to create the specimen that behaves like the real material, the same procedure as introduced in the authors' previous work follows [1]. Figure 1 shows the flow chart which indicates how this procedure works.

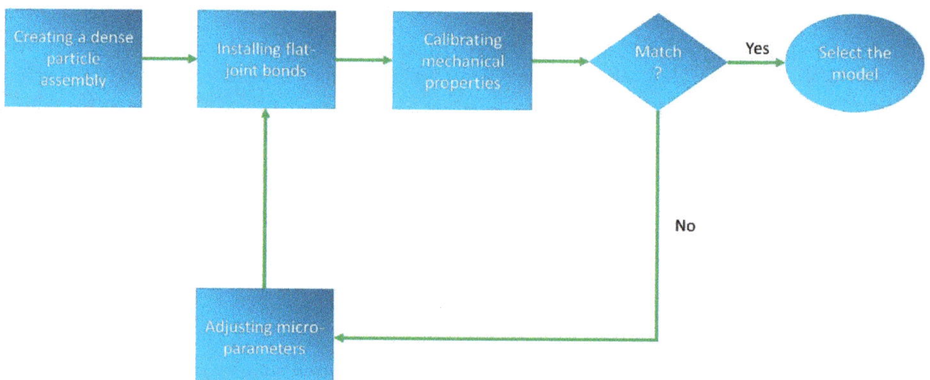

Figure 1. Flow chart of the specimen creation procedure.

The first step is to create a dense particle assembly. In this step, macro-parameters including the specimen density, sample dimensions, and particle radii are defined. Then, the particles are bonded based on the flat-joint contact model. The micro-parameters are defined in this step which contain the particle contact stiffness, particle stiffness ratio, particle friction coefficient, particle damping coefficient, bond shear strength, bond normal strength, and friction angle. The macro-parameters and micro-parameters used in this study are shown in Table 1. The third step is to match the mechanical properties of the specimen. Uniaxial tensile and compressive tests are simulated; four mechanical properties are matched in this step which contains elastic modulus, tensile strength, compressive strength, and poisson's ratio. More details of the simulation procedure are provided elsewhere [1]. Through a series of adjustments, the DEM model is calibrated to match the soda-lime glass properties as listed in Table 2.

Table 1. Macro- and micro-parameters of the synthetic specimen.

Micro-Parameters	Description	Value
ρ	Ball density (kg/m^3)	2.4×10^3
H	Sample height (m)	1.0×10^{-3}
W	Sample width (m)	2.0×10^{-3}
R_{min}	Minimum ball radius (m)	2.5×10^{-6}
R_{max}	Maximum ball radius (m)	5.0×10^{-6}
Micro-Parameters	**Description**	**Value**
E_c	Ball–ball contact modulus (Pa)	8.1×10^{10}
k_n/k_s	Ball stiffness ratio	4.3
$\bar{E}_c$	Flat-joint bond modulus (Pa)	8.1×10^{10}
$\bar{k}_n/\bar{k}_s$	Flat-joint bond stiffness ratio	4.3
μ	Ball friction coefficient	0.577
$\bar{\sigma}_c$	Flat-joint normal strength (Pa)	1.69×10^8
$\bar{\tau}_c$	Flat-joint shear strength (Pa)	1.85×10^8
ϕ	Friction angle (degree)	25.0

Table 2. Comparison of properties between the discrete element method (DEM) model and soda-lime glass.

Mechanical Property	Elastic Modulus E (GPa)	Tensile Strength σ_t (MPa)	Compressive Strength σ_c (MPa)	Poisson's Ratio v
Soda-lime glass	71	41	330	0.23
DEM Model	71.1	42.6	332	0.228

2.2. Model Validation

After the material model is calibrated, orthogonal cutting of the same synthetic material is simulated. Figure 2a shows the model geometry and boundary conditions. The workpiece is 2 mm in length and 1 mm in height, which contains a total of 9885 particles. The particles marked as red are fixed to simulate the boundary conditions. The cutting tool is modeled as a rigid body which has a rake angle of −15° and clearance angle of 15°. The depth of cut is 0.1 mm and the cutting speed is 4 mm/s.

The orthogonal machining experimental setup is constructed in order to validate the simulation model. As shown in Figure 2b, a vertical Bridgeport milling machine with a cutting tool fixed on the locked vertical spindle column is used to cut the sample. The soda-lime glass sample is fixed on the horizontal carriage which is mounted on the Kistler three-component dynamometer (Kistler Instrument Corp, Novi, MI, USA). A Kistler dual-mode charge amplifier amplifies the cutting force signals, which are measured by the dynamometer with the sampling rate of 200 Hz. LabView (National Instrument, Austin, TX, USA) is used to control the computer data acquisition system. The workpiece, fixture, and dynamometer are clamped on the movable carriage of the milling machine, and the carriage feeds the workpiece to the cutting tool. The cutting tool is a $16 \times 16 \times 6$ mm^3 square ceramic insert made of alumina and is mounted on the tool holder tilted to attain a negative 15° rake angle. The cutting conditions for both the simulation and experiments are given in Table 3.

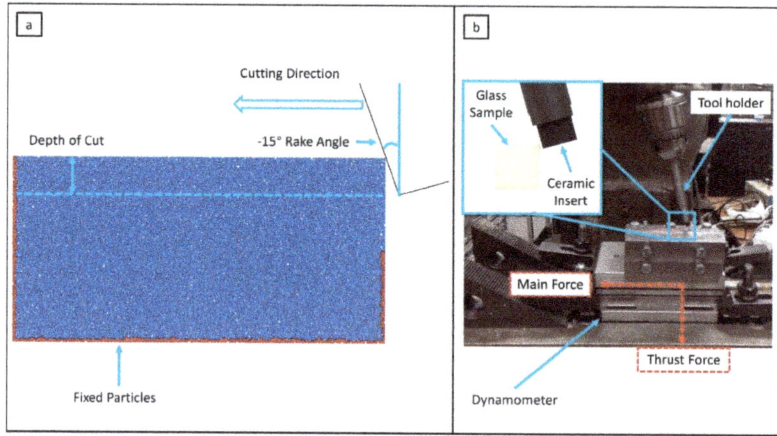

Figure 2. (**a**) DEM model of the cutting simulation, (**b**) experimental setup of the orthogonal machining test.

Table 3. Cutting conditions for the experiments.

Parameter	Description	Simulation	Experiment
V (mm/s)	Cutting speed	4	4
t (mm)	Depth of cut	0.1	0.1
α (degree)	Rake angle	−15	−15
L (mm)	Length of cut	2	25.4

In the chip formation image as shown in Figure 3a, the green, short lines denote broken bonds between the particles which are caused by shear failure; the red, short lines are also broken bonds, but they are caused by tensile failure. Broken bonds are considered as random cracks. It can be seen that many broken bonds are connected to each other and continue propagating to a deeper region, which is the formation of subsurface cracks. Some of the subsurface cracks can even propagate a few hundred micrometers in distance. Compared to the optical images in Figure 3b, similar subsurface cracks can be clearly observed in the cutting experiments.

In addition, the chips are recorded for both the cutting simulation and experiments. In the simulation, the chips are formed due to random propagation of broken bonds. The bonded particles are separated by broken bonds into smaller segments with different shapes. The chips from the cutting experiments are also collected and examined. It can be seen clearly that the shapes and dimension of those chips are very similar to those from the simulation.

The cutting forces in both the horizontal (main) and vertical (thrust) directions are recorded during the simulation and experiments. It can be seen that numerous force peaks exist for both the simulation and experimental workpieces, which are caused by the initiation and propagation of cracks due to bond breakage between particles. This behavior is typical for brittle material removal processes, characterized by random peaks and valleys which correspond to force build-up followed by sudden fracture occurrence. It is found that the force magnitude for the simulation workpiece is quite similar to the experimental one. In order to reduce the influence of particle arrangement during the simulation, the specimens are generated with different random particle arrangements. Both simulation and experiments are repeated three times, the average cutting forces for each replication are listed in Table 4.

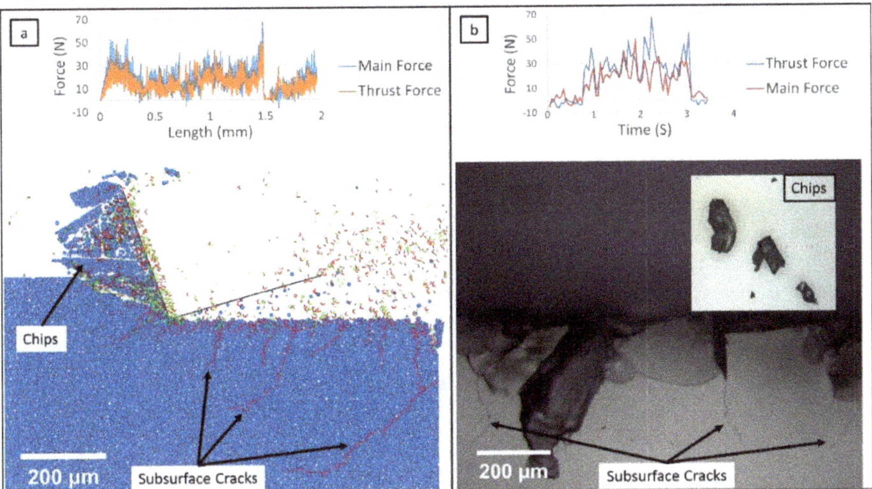

Figure 3. (**a**) DEM model of the cutting simulation and recorded forces, (**b**) optical images of the machined area of the soda-lime glass sample and recorded forces.

Table 4. Average cutting forces for each experiment.

Depth of Cut (mm)	Replications	Average Cutting Force (N)			
		Experimental Workpiece		Simulation Workpiece	
		Main	Thrust	Main	Thrust
	1	17.8	14.1	15.4	10.1
0.1	2	15.0	12.5	16.4	11.6
	3	16.7	14.0	17.5	11.8

In order to compare the simulation and experimental results, the average forces for the three replications are taken and plotted in Figure 4. The main forces for both the simulation and experiments are almost the same, but the thrust force from the simulation is 17% less than that from the experiment results. Due to the brittle feature of glass material, this difference is considered acceptable. Hence, the DEM model is validated through this process.

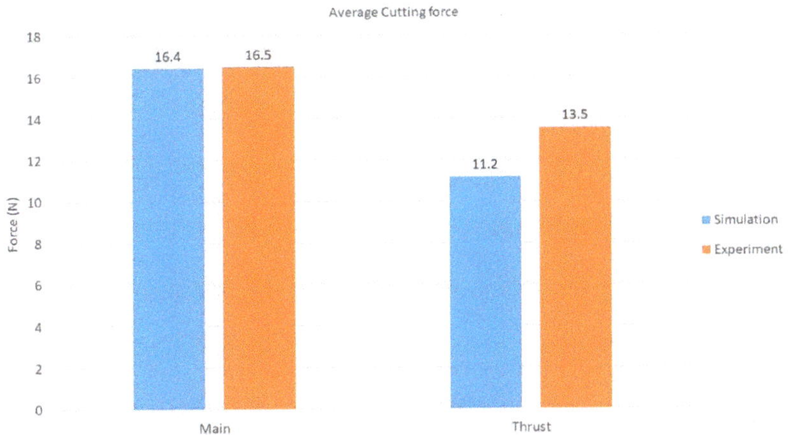

Figure 4. Cutting force comparison for the simulation and experimental results.

3. Surface Roughness Prediction

Surface roughness plays an important role in determining product quality and in most cases is a technical requirement for mechanical products. The functional behavior of a part is highly dependent on the desired surface quality. Machining simulations are well studied through the years, however, predicting surface roughness is much more difficult than predicting cutting forces due to modeling complexities. Since the discrete element method models the workpiece as bonded particles, the position of each particle can be tracked through the whole simulation process. Based on this idea, an algorithm has been developed to simulate the surface roughness of the machined workpiece.

The first step is to identify the particles which are not separated from the main workpiece. As can be seen from Figure 5a, there are numerous broken bonds generated after the cutting is done. Some of the particles are ejected from the main workpiece due to broken bonds, and some of them are still attached to the main workpiece but are not bonded to it anymore. Under this circumstance, the particles which are separated from the main workpiece need to be filtered out. A critical displacement of 10^{-7} m is used as a criterion to decide whether a particle remains on the surface. As shown in Figure 5b, the total displacement of particle P1 is larger than the critical value, so it is not considered as part of the main workpiece. Although particle P2 is not bonded to the workpiece, the total displacement of P2 is within the critical value, so it is still considered as part of the main workpiece. Since P2 is located at the top surface, P2 belongs to the surface particles.

After filtering out the particles which belong to the main workpiece, those that form the top surface of the machined part are identified. This is done by first dividing the cutting distance into finite intervals, as shown in Figure 5b. Then, the highest particle within each interval is selected as the surface particle. Care should be taken to select the length of these intervals. If it is too large, some surface particles will be missing; if it is too small, unwanted particles will be generated and interfere with the actual surface profile.

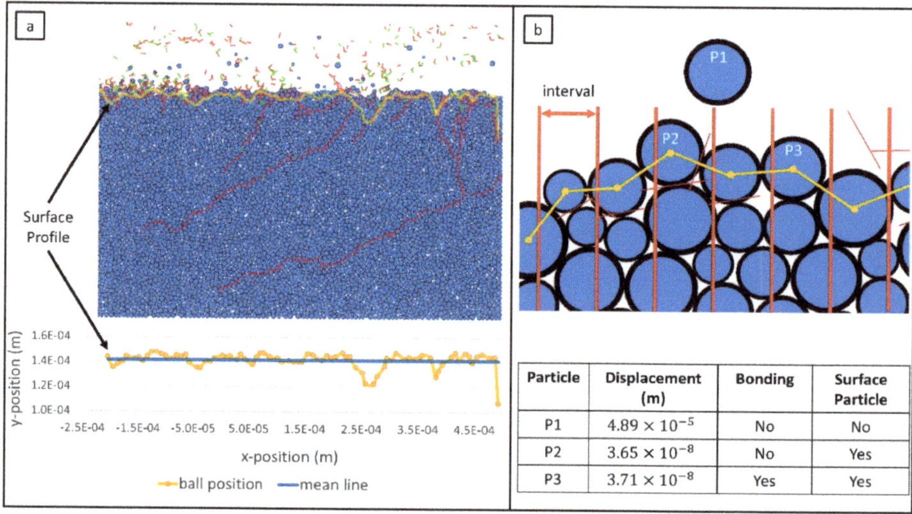

Particle	Displacement (m)	Bonding	Surface Particle
P1	4.89×10^{-5}	No	No
P2	3.65×10^{-8}	No	Yes
P3	3.71×10^{-8}	Yes	Yes

Figure 5. (a) Surface profile of a machined sample by DEM simulation, (b) surface particles.

The common measure for surface roughness is known as *Ra* [13], which is the arithmetical mean deviation of the surface profile governed by:

$$Ra = \frac{1}{n} \sum_{i=1}^{n} |y_i - m|$$

where n is the number or total particles, y_i is the y-position of the particle, m is the mean value of the y-positions for all the surface particles which is expressed as:

$$m = \frac{1}{n}\sum_{i=1}^{n} y_i$$

The *Ra* value of the surface profile as shown in Figure 5a is found to be 4.25 μm.

4. Effects of Seed Crack Types on Cutting Performance

The orthogonal cutting with different seed cracks is simulated and each simulation is repeated three times using specimens generated with different random particle arrangements to reduce the influence of particle arrangement. Four different types of seed crack oriented at the angle (θ) 0°, 45°, 90°, and 135° with the horizontal direction are shown in Figure 6a–d. The height of the seed cracks (h) is 100 μm under the surface, the width of each seed crack (w) is around 10 μm, and the distance between adjacent seed cracks (d) is 200 μm. The cutting conditions are shown in Table 5. The cutting speed is set at 1 m/s, the depth of cut is 0.1 mm, the rake angle is −15°, and the width of cut is 2 mm. Figure 7a–d shows images for different conditions taken during the cutting simulation. The green particles indicate the fixed boundaries. During the simulation, the main (horizontal direction) and thrust (vertical direction) cutting forces are recorded; random crack numbers and the Ra values of surface roughness are recorded as well. Table 6 shows the results for each cutting condition.

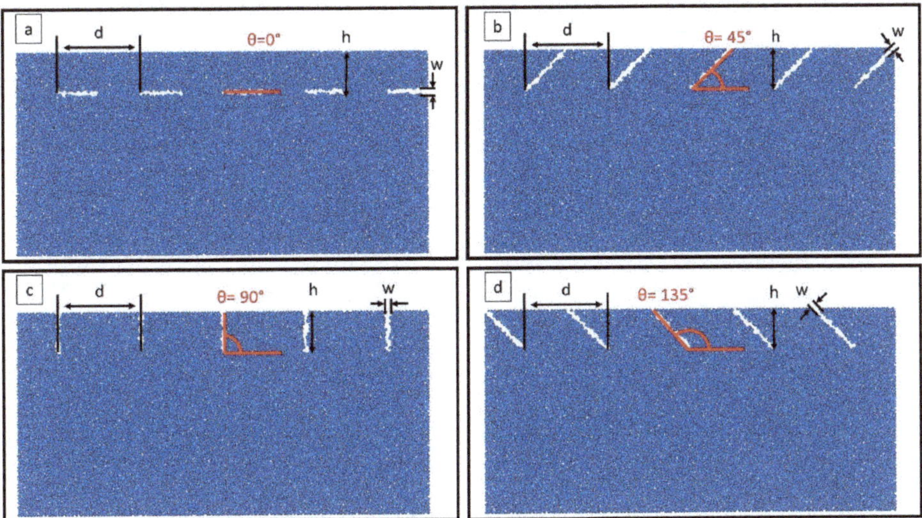

Figure 6. Synthetic workpieces with seed cracks of different angles: (**a**) 0°, (**b**) 45°, (**c**) 90°, and (**d**) 135°.

Table 5. Cutting conditions for the simulation.

Parameter	Description	Value
V (m/s)	Cutting speed	1
t (mm)	Depth of cut	0.1
α (degree)	Rake angle	−15
L (mm)	Length of cut	2

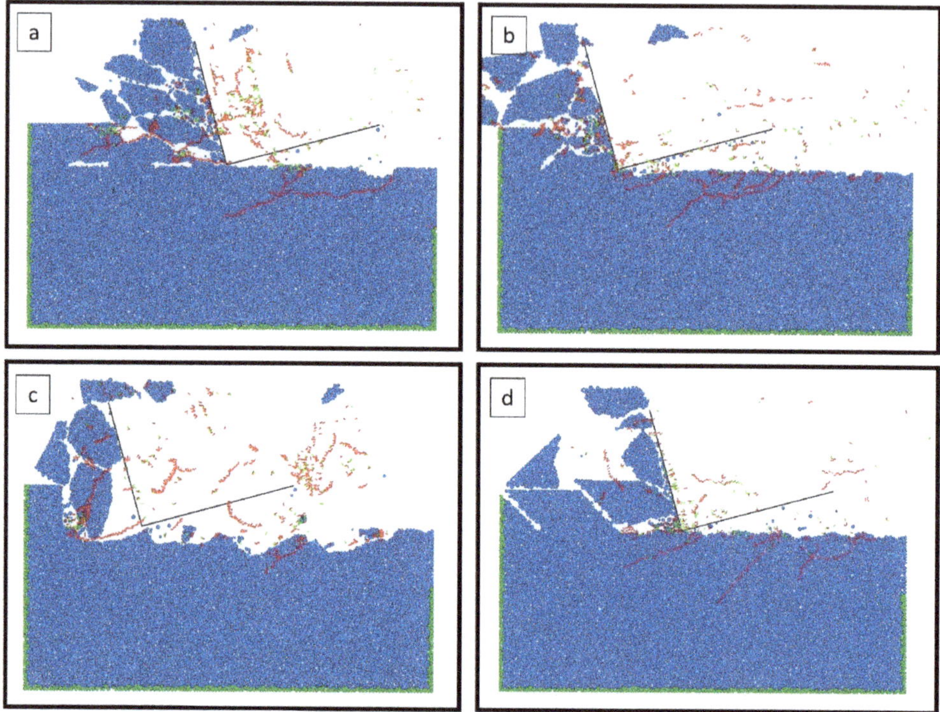

Figure 7. Simulation images of machined workpiece with seed cracks of different angles: (**a**) 0°, (**b**) 45°, (**c**) 90°, and (**d**) 135°.

Table 6. Simulation results for each condition.

Angle of Seed Cracks	Replications	Average Cutting Force (N)		Random Cracks	Surface Roughness *Ra* (μm)
		Main	Thrust		
0°	1	2.58	2.14	2893	29.71
	2	3.59	2.33	3424	6.70
	3	3.72	2.05	4212	9.38
45°	1	1.80	1.75	2268	9.01
	2	1.43	1.46	2010	10.21
	3	1.46	1.73	1814	21.90
90°	1	2.48	1.86	2735	28.30
	2	3.48	1.95	1577	15.52
	3	2.54	2.00	3504	20.41
135°	1	1.79	1.11	2110	8.29
	2	1.98	1.71	1674	20.28
	3	1.21	1.53	2223	9.52

In order to compare the simulation results for each condition, the average values for the cutting force, random cracks, and surface roughness over the three replications are taken and plotted in Figure 8a–c. In general, compared to the untreated samples, cutting the treated samples with seed cracks can greatly reduce the cutting force, random cracks, and surface roughness. As can be seen, the cutting forces are reduced by 80%–90%, the random cracks are reduced by 54%–74%, and the surface roughness is reduced by 10%–47%.

Comparing these four different seed cracks, the 45° and 135° conditions are better than the 0° and 90° conditions in general. Based on the results for cutting forces, random cracks, and surface roughness, the 45° and 135° conditions give the better performance than the 0° and 90° conditions. Between the 45° and 135° conditions, the cutting simulations predict that the overall performance under the 135° seed crack condition is better.

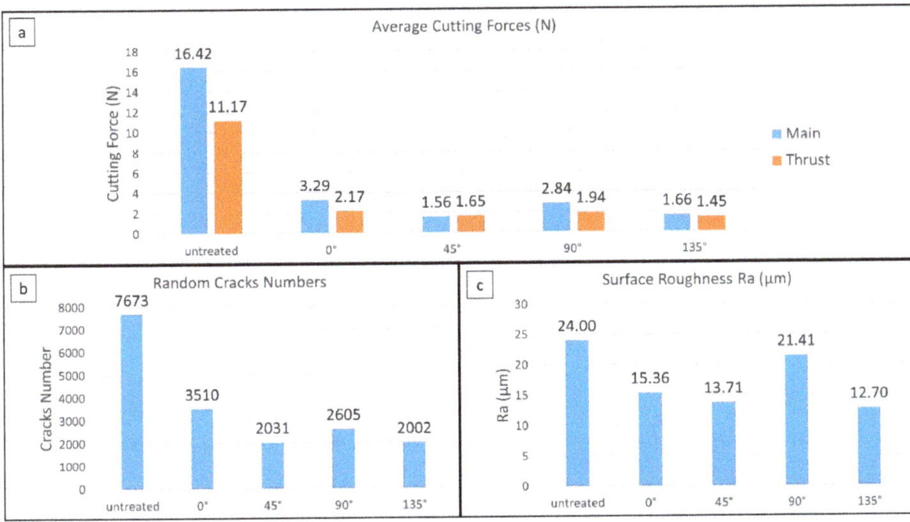

Figure 8. The simulation results for four different conditions: (**a**) Average cutting forces, (**b**) average random cracks, and (**c**) average surface roughness.

5. Conclusions

In this paper, the effects of seed cracks are studied and predicted through the orthogonal cutting simulation with PFC2D based on the discrete element method. The results of cutting forces, random cracks, and surface roughness are analyzed. The following conclusions are obtained from this study:

- The simulation results show that the cutting forces can be greatly reduced by cutting workpieces with seed cracks.
- When the seed cracks are orientated at 45° and 135°, the cutting forces can be minimized.
- When the seed cracks are orientated at 135°, the random cracks and surface roughness are able to be minimized.

Author Contributions: G.Y. and S.L. conceived and conducted the simulation; H.A. and G.Y. performed the experiments; G.Y. analyzed the data; and G.Y. and S.L. wrote the paper. All authors have read and agreed to the published version of the manuscript.

Funding: This research was funded by the National Science Foundation under grant no. CMMI-1537846.

References

1. Yang, G.; Alkotami, H.; Lei, S. Discrete Element Simulation of Orthogonal Machining of Soda-Lime Glass. *J. Manuf. Mater. Process.* **2018**, *2*, 10.
2. Cundall, P.A.; Strack, O.D.L. A discrete numerical model for granular assemblies. *Geotechnique* **1979**, *29*, 47–65. [CrossRef]
3. Cheng, Y.; Nakata, Y.; Bolton, M.J.G. Discrete element simulation of crushable soil. *Geotechnique* **2003**, *53*, 633–641. [CrossRef]

4. Langston, P.A.; Tüzün, U.; Heyes, D.M. Discrete element simulation of granular flow in 2D and 3D hoppers: Dependence of discharge rate and wall stress on particle interactions. *Chem. Eng. Sci.* **1995**, *50*, 967–987. [CrossRef]

5. Tang, C.-L.; Hu, J.-C.; Lin, M.-L.; Angelier, J.; Lu, C.-Y.; Chan, Y.-C.; Chu, H.-T. The Tsaoling landslide triggered by the Chi-Chi earthquake, Taiwan: Insights from a discrete element simulation. *Eng. Geol.* **2009**, *106*, 1–19. [CrossRef]

6. Huang, H.-Y.; Lecampion, B.; Detournay, E. Discrete element modeling of tool-rock interaction I: Rock cutting. *Int. J. Numer. Anal. Methods Geomech.* **2013**, *37*, 1913–1929. [CrossRef]

7. Shen, X.; Lei, S. Distinct Element Simulation of Laser Assisted Machining of Silicon Nitride Ceramics: Surface/Subsurface Cracks and Damage. In Proceedings of the ASME 2005 International Mechanical Engineering Congress and Exposition, Orlando, FL, USA, 5–11 November 2005; pp. 1267–1274.

8. Iliescu, D.; Gehin, D.; Iordanoff, I.; Girot, F.; Gutiérrez, M.E. A discrete element method for the simulation of CFRP cutting. *Compos. Sci. Technol.* **2010**, *70*, 73–80. [CrossRef]

9. El-Hofy, H.A.-G. *Advanced Machining Processes: Nontraditional and Hybrid Machining Processes*; McGraw Hill Professional: Two Penn Plaza, NY, USA, 2005.

10. Brehl, D.E.; Dow, T. Review of vibration-Assisted machining. *Precis. Eng.* **2008**, *32*, 153–172. [CrossRef]

11. Choi, J.; Jeon, B.; Kim, B.H. Chemical-Assisted ultrasonic machining of glass. *J. Mater. Process. Technol.* **2007**, *191*, 153–156. [CrossRef]

12. Shanmugam, N.; Yu, X.; Alkotami, H.; Devin, G.; Lei, S. Machining of Transparent Brittle Material Assisted by Laser-Induced Seed Cracks. In Proceedings of the ASME 2016 11th International Manufacturing Science and Engineering Conference, Blacksburg, VA, USA, 27 June–1 July 2016; p. V001T002A019.

13. Black, J.T.; Kohser, R.A.; De Garmo, E.P. *Degarmos Materials and Processes in Manufacturing*; John Wiley & Sons, Inc.: Hoboken, NJ, USA, 2018.

Journal of
*Manufacturing and
Materials Processing*

Article

Effect of Post Treatment on the Microstructure, Surface Roughness and Residual Stress Regarding the Fatigue Strength of Selectively Laser Melted AlSi10Mg Structures

Wolfgang Schneller [1,*], Martin Leitner [1], Sebastian Pomberger [2], Sebastian Springer [1], Florian Beter [1] and Florian Grün [1]

[1] Montanuniversität Leoben, Department Product Engineering, Chair of Mechanical Engineering, Franz-Josef-Straße 18, 8700 Leoben, Austria; martin.leitner@unileoben.ac.at (M.L.); sebastian.springer@unileoben.ac.at (S.S.); florian.beter@unileoben.ac.at (F.B.); florian.gruen@unileoben.ac.at (F.G.)
[2] Christian Doppler Laboratory for Manufacturing Process Based Component Design, Montanuniversität Leoben, Franz-Josef-Straße 18, 8700 Leoben, Austria; sebastian.pomberger@unileoben.ac.at
* Correspondence: wolfgang.schneller@unileoben.ac.at

Received: 24 September 2019; Accepted: 14 October 2019; Published: 16 October 2019

Abstract: This paper focusses on the effect of hot isostatic pressing (HIP) and a solution annealing post treatment on the fatigue strength of selectively laser melted (SLM) AlSi10Mg structures. The aim of this work is to assess the effect of the unprocessed (as-built) surface and residual stresses, regarding the fatigue behaviour for each condition. The surface roughness of unprocessed specimens is evaluated based on digital light optical microscopy and subsequent three-dimensional image post processing. To holistically characterize contributing factors to the fatigue strength, the axial surface residual stress of all specimens with unprocessed surfaces is measured using X-ray diffraction. Furthermore, the in-depth residual stress distribution of selected samples is analyzed. The fatigue strength is evaluated by tension-compression high-cycle fatigue tests under a load stress ratio of R = −1. For the machined specimens, intrinsic defects like pores or intermetallic phases are identified as the failure origin. Regarding the unprocessed test series, surface features cause the failures that correspond to significantly reduced cyclic material properties of approximately −60% referring to machined ones. There are beneficial effects on the surface roughness and residual stresses evoked due to the post treatments. Considering the aforementioned influencing factors, this study provides a fatigue assessment of the mentioned conditions of the investigated Al-material.

Keywords: fatigue; SLM; AlSi10Mg; post treatment; residual stress; surface roughness

1. Introduction

Selective laser melting (SLM) enables the manufacturability of complexly shaped and topographically optimized components. Additive manufacturing (AM) is contemplated to find significant application in demanding fields such as automotive, aviation and biomedical engineering [1–5]. Particularly in complex structures, post built machining is not always possible; hence, it is of upmost importance to investigate the influence of the unprocessed surface on the fatigue strength in conjunction with the effect of subsequent post treatments [6,7]. It is estimated that about 90% of all engineering failures are caused by fatigue-related damage mechanisms [8,9]. Along with Ni-based alloys, stainless steel and titanium, aluminum alloys, AlSi10Mg is especially a very commonly used material for powder-bed based AM and therefore causes the necessity of a proper as well as safe assessment of the material qualification regarding fatigue [10]. Current studies on stainless and tool steels as well as titanimum alloys deal with the importance of surface quality, process parameters as well as post treatments and possible reasons for defects formations. For example, powder defects, insufficient energy and consequent partially melted powder particles or material vaporization impact static and cyclic material properties [11–16]. Additionally, the manufacturability of lattice structure by AM provides huge potential in terms of lightweight design and is subject to many research works. The interaction between the building direction, microstructure, and crack propagation is discussed in [11]. The microstructure is found to have great influence on the fatigue crack morphology and crack deflection effects. Fatigue crack initiation and the propagation rate play a major role in fatigue properties, whereby it is found that initiation is strongly linked with the surface roughness and the crack propagation rate with the microstructure and stress level [17]. Among others, hot isostatic pressing (HIP) and solution annealing (T6) are two common procedures to enhance material properties [18–20]. Given the fact that HIP leads to a reduction of the volume fraction of porosity and improved fatigue resistance for sand-casted aluminum components, an according HIP treatment may be beneficial to AM parts as well [21–23]. SLM structures generally exhibit an extremely fine microstructure due to high cooling rates [24]. A heat treatment above the solubility temperature of AlSi10Mg causes microstructural coarsening, since grain boundaries are dissolved as well as the precipitation of second phase particles [9,25–27]. These microstructural changes result in reduced fatigue properties, and therefore demand a subsequent age hardening process in order to counteract those unfavourable effects [28]. The exact post treatment parameters are set up incorporating the knowledge of the specimen manufacturer. The influence of the post treatments is further investigated in terms of the surface roughness and residual stresses. The fatigue strength of engineering components is decreased with increasing surface roughness. Elevated surface roughness tends to generate stress concentration factors and favors failure initiation [6]. In this study, the effect of the unprocessed surface is investigated and described using a notch effect factor referring to a machined condition [29,30]. The applicability of an endurance limit reducing factor is researched and validated with experimental results. The impact of residual stresses on the fatigue strength is studied as well within this work. It is of utmost importance to holistically assess material qualification, since a present residual stress state can significantly alter the stress condition at the failure initiating imperfection [31,32]. A post treatment also influences the residual stress condition in great measure. Neglecting residual stresses may lead to non-conservative designing of components, which is the reason for the conducted research work. It is of technical and economical relevance to investigate the influence of residual stresses and enhance existing concepts to properly as well as safe assess material qualifications regarding fatigue. This study provides a method how to assess the impact of surface features under consideration of residual stresses acting as mean stresses. The authors propose an approach to account for residual stresses in fatigue design and furthermore look at notch effects due to surface roughness independently, which allows a differentiated assessment of roughness features and residual stress effects.

2. Materials and Methods

Three different post treatment conditions are the subject of this work. Therefore, it was necessary to clearly distinguish between the test series. The following enumeration clarifies the abbreviations used in the present study and provides the applied treatments for each condition. A detailed description of the respective routines is given in Table 1. The first column refers to the treatment, followed by temperature, pressure and time, which provides information about the minimum holding time of the respective treatment. The exact post treatment parameter is defined incorporating the knowledge of the specimen manufacturer, aiming to enhance material properties. For this reason, the used parameter sets are classified:

- Test series "AB": As-built condition (no post treatment applied),
- Test series "HIP": Hot isostatic pressing + age hardening,
- Test series "SA": Solution annealing + age hardening.

Table 1. Parameter of subsequent post treatments.

Treatment	T (°C)	P (MPa)	Time (h)
Hot isostatic pressing	above 500	above 100	2
Solution annealing	above 500	-	6
Age hardening	below 200	-	7

In order to quantify the impact of the surface roughness, each of the above-mentioned test series (AB, HIP and SA) consisted of two batches—one lot exhibiting a machined and polished surface—denoted as "M", and a second set of specimens in as-built (not machined) surface condition—denoted as unprocessed "UP". Therefore, in total, six test series were investigated. There were nine specimens that exhibited a polished surface and five specimens with unprocessed surfaces manufactured for each condition. The abbreviation for the surface condition was added before the post treatment e.g., M-HIP means machined surface and HIP treated or UP-SA stands for unprocessed surface and solution annealing.

The used AlSi10Mg powder for specimen manufacturing showed the chemical composition given by the powder manufacturer in Table 2 [33]. According to manufacturer specifications, the material corresponds to the standard DIN EN 1706:2010 [34].

Table 2. Chemical composition of the AM powder in weight %.

Material	Si	Fe	Cu	Mn	Mg	Al
AlSi10Mg	9.0–11.0	0.55	0.05	0.45	0.20–0.45	Balance

All specimens were built in a vertical direction on an EOS M290 system, using a Yb fiber laser with a power of 400 W. The beam diameter is set to 100 μm. The standard parameter set provided by EOS is used for printing. To ensure all surface-related effects are eliminated for the investigation of the machined conditions, a respective number of specimens is manufactured with a certain machining allowance to subsequently remove the boundary layer. Following the manufacturing process, the respective post treatment was applied. Afterwards, the specific specimens for the machined test series were processed to the geometry by turning and polishing, shown in Figure 1. The geometry of the specimen corresponds to no standard but is designed to minimize the stress concentration within the testing section caused by the narrowing shape. A numerical analysis reveals a maximum principal stress concentration of $K_t = 1.045$, hence 4.5% at the thinnest point. The same specimen geometry and manufacturing parameter are used for previous work already published by the authors in [35].

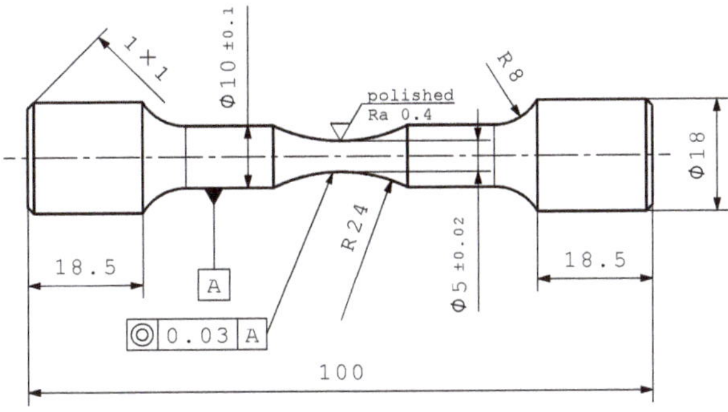

Figure 1. Specimen geometry for high-cycle fatigue testing [35].

2.1. SEM Investigation

To characterize the impact of the respective post treatment on the microstructure, backscatter-SEM images of microsections were taken with a Carl Zeiss EVO MA 15 microscope in accordance with [36]. Both post treatments were conducted above the solubility temperature of the investigated material [37–39]. It is mentioned that the solution temperature of the cast alloy is above 450 °C, and therefore a subsequent age hardening at low temperatures leaves the microstructural evolution unchanged [40].

2.2. High Cycle Fatigue Assessment

For all test series, a modified staircase test method was utilized [41]. The high-cycle fatigue testing was carried out under a load stress ratio of R = −1 on an RUMUL Mikrotron resonant testing rig. The test frequency was in the region of 106 Hz. Specimens were gripped with collets at both ends. The test was aborted when total fracture occurred, or the run-out criterion of 1E7 load cycles was reached. In order to generate more data within the finite life region, conservatively not ruling out the possibility of pre-damaging at load levels below the fatigue limit, run-outs were reinserted [42]. In the following work, selected results referring to the AB and HIP conditions have been partially published within preliminary studies in [35]. All given stress values were normalized to the nominal ultimate tensile strength (UTS) of the base material without any post treatment, given by the powder manufacturer [33]. The fatigue strength at 1E7 load-cycles for a survival probability of 50% (σ_f) was statistically determined by applying the $\arcsin\sqrt{P}$-transformation, described in [43]. The assessment of the S/N-curve within the finite life region was done utilizing the ASTM E739 standard [44].

Mean stresses impact the fatigue strength whereby the endurance limit is decreased with growing mean stresses such as static loads along with cyclic loading [45]. The effect is usually depicted as fatigue strength amplitude plotted over mean stress. A large number of concepts have been developed in order to predict the fatigue strength for different mean stress states [46,47]. Two models, one according to Gerber [48] and another one developed by Dietman [49] were utilized within this work to consider a certain mean stress state caused by residual stresses and its impact on fatigue. Equations (1) and (2) serve as two models to correct the endured stress amplitude dependent on the present residual stress state. Both required the ultimate tensile strength σ_{uts} for the respective condition, which was provided by the specimen manufacturer. The parabolic Gerber concept as well as the empirical Dietmann equation showed high statistical correlation with experimental data, which is why those two models were applied. In the following, $\sigma_{a(-1)}$ stands for the stress amplitude at a load stress ratio of R = −1, and σ_m refers to the present mean stress. Considering this, the endurable stress amplitude σ_a, at a certain mean stress, can be estimated:

$$\sigma_a = \sigma_{a(-1)}\left[1 - \left(\frac{\sigma_m}{\sigma_{uts}}\right)^2\right], \tag{1}$$

$$\sigma_a = \sigma_{a(-1)}\sqrt{1 - \frac{\sigma_m}{\sigma_{uts}}}. \tag{2}$$

2.3. Residual Stress Measurement Methodology

The holistic characterization of contributing factors to the fatigue strength causes the necessity to assess the residual stress state [31,32], especially in regard to the building process [50,51]. The analysis was performed with X-ray diffraction using an X-RAYBOT from MRX-RAYS, located in Brumath, France. A psi-mounting configuration with Cr-Kα radiation was used along with a collimator size of 2 mm in diameter. The evaluation was based on the $2\theta - \sin^2(\psi)$ method. The measurement setup was according to the ASTM E915-96 standard [52]. The exposure time was set to 30 s for each increment, opting for 25 ψ-increments, with a tilting angle of the X-ray tube from −40° to +40°. The measurement procedure corresponds to the ASTM E2860-12 standard [53]. The residual stress analysis is performed on all unprocessed specimens to avoid falsifying of the results due to influences of machining. Since the fatigue strength at 1E7 load cycles is of interest, one should be aware of a possible depletion of residual stress under tensile loading. For this reason, the validation of the cyclic stability of residual stresses is necessary in order to ensure the usability of the measured stresses in following work. Therefore, in situ residual stress measurements were conducted while fatigue testing. For the assessment of the cyclic stability of the present residual stresses, the fatigue testing was stopped, residual stresses were measured, and, afterwards, the testing is continued. In order to avoid falsifying of the results, the specimen remains clamped in the testing rig.

2.4. Surface Roughness Evaluation

An engineering approach to characterize the reduction of the fatigue strength due to the surface roughness includes the maximum depth of roughness valleys as well as the roughness valley radius. Based on a concept of Peterson, the unprocessed surface, exhibiting micro notches due to the building process, was characterized. Considering the localized stress concentration of such features, the consequent reduction of fatigue properties can be described by the notch effect factor K_t; see Equation (3) [30]. This approximate solution for a shallow, assumed ideal elliptical notch, is only a function of the notch depth and radius of the curvature. Therefore, this concept incorporated the maximum surface deviation S_t and the notch root radius ρ. Based on recommendations by the author, the support effect was not taken into account and set to $n = 1$ due to a conservative approach; for this reason, K_t equals K_f. This concept finds application within this study to predict the reduced endurable stress amplitude of the unprocessed specimens, beginning with the fatigue strength of the machined ones, respectively, in mean (residual) stress free state:

$$K_t = 1 + 2\sqrt{\frac{S_t}{\rho}}. \tag{3}$$

Utilizing a light optical microscope and three-dimensional image processing, it was possible to determine the average maximum surface deviation (S_t) in a non-destructive way [54], shown in Figure 2. Since the specimen geometry is round and additionally possesses a curvature within the testing area, proper filtering of the captured surface topography is necessary. In a first step, the round specimen was partitioned into 12 sections that are individually captured and represent the entire surface. Exemplary, Figure 3a pictures the primary profile, respectively the geometrical structure of one surface segment, detected by the digital optical microscope. The thereby generated three-dimensional datasets were processed within a user-defined routine, as described in [55]. By means of a second order robust Gaussian regression filter, the roughness profile is calculated applying a cut-off wavelength λ_c of 2.5 mm. The cut-off length was chosen as recommended by the authors in [55]. This results in the waviness profile as pictured in Figure 3b and the associated roughness profile, see Figure 3c, of the exemplified surface segment. The roughness profile now entirely reflects the surface topography as the waviness profile corresponds to the specimen geometry, respectively form. After areal roughness calculation, the evaluated area is separated into sub-areas, 1×1 mm^2 in size by means of the routine and plotted onto the measured surface image. An exemplary roughness map of the areal roughness parameter S_t is shown in Figure 2. Yellow areas mark high roughness values, and blue areas mark low ones. Due to that, not only can local areal roughness parameters be linked to surface topography properties, such as notch depth, but also information about the location of the structures are gained.

Figure 2. Exemplary surface roughness map.

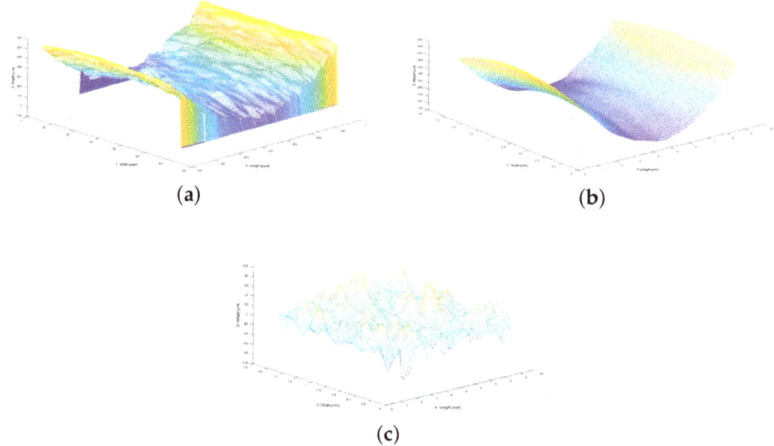

(a)

(b)

(c)

Figure 3. Surface roughness evaluation process. (**a**) Primary surface profile. (**b**) Waviness surface profile. (**c**) Surface roughness profile.

3. Results

3.1. Microstructural Analysis

In the untreated condition, see Figure 4a, one can identify pores and grain boundaries, also detected in [56]. The post treated conditions differ from the as-built condition, as significant changes in the microstructure are detected. Grain boundaries are no longer clearly visible, and precipitates are formed within the microstructure. This is observed for both post treatments; see Figures 4b and 5. By virtue of the heat influence, the post treatment causes melt pool boundary softening, implying microstructural evolution and precipitation [57]. Additionally, the porosity and the maximum extension of pores are significantly decreased for the HIP condition, also detected in [58] and published within previous work on this topic in [35].

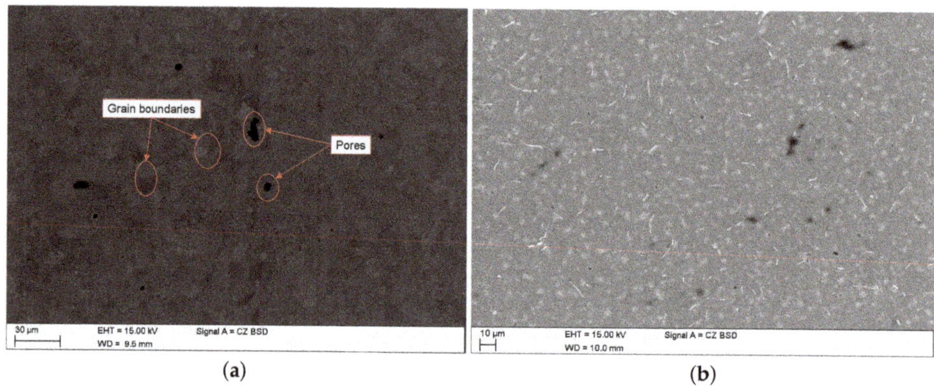

(a)

(b)

Figure 4. Microstructural analysis. (**a**) Microstructure of the AB condition [35]. (**b**) Microstructure of the SA condition.

The changes to the microstructure found in the conditions with a heat-treatment above 500 °C are investigated in detail. Iron-rich precipitates and silicon agglomerations are detected; compare [59].

These microstructural features are also found in [27] for both the HIP and SA conditions. A performed EDX-analysis on a Fe-rich precipitate, the spot marked as 'a' in Figure 5, shows a chemical composition ($Al_{70.24}Si_{15.24}Fe_{14.32}$) that calculates to $Al_5Si_{1.1}Fe_{1.02}$ and is similar to the β-phase Al_5SiFe, reported and found in [60–62]. Due to the elevated temperature above the solubility temperature, silicon crystals are precipitated at the grain boundaries which grow to their respective size throughout the subsequent annealing [37,38,63]. An analysis at spot 'c' confirms the labelled agglomerations as Si-particles that are well reported in [64,65]. The detected microstructural features decelerate the long crack growth. The crack front interferes with these microstructural features, and the propagation is obstructed and forced to change its direction, whereby the overall resistance against fatigue crack growth is enhanced. The improved resistance against crack propagation is attributed to deflection and energy dissipation at the crack tip [25,66]. Within this study, this microstructural behavior is observed for the HIP and the SA condition; compare [35]. After the post treatment, the base material in area 'b' shows a chemical composition of $Al_{94.27}Si_{5.73}$, which differentiates to the as-built matrix due to precipitation.

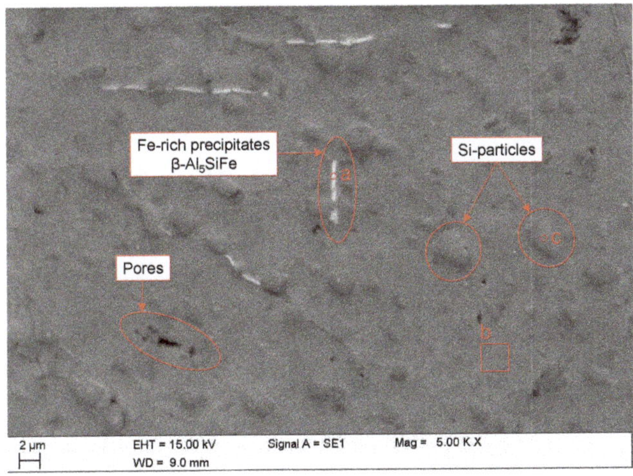

Figure 5. Microstructure in post treated condition including EDX analysis.

3.2. Residual Stress Measurement

3.2.1. Surface Residual Stresses and Cyclic Stability

For the unprocessed condition, it is highly necessary to know the residual stresses at the surface, since this is the location of the failure origin, and the condition within the failure initiation area is essential. The interaction between surface condition, residual stresses, and, furthermore, the microstructure as well as understanding the importance of their codependency is also reported in [67,68]. To ensure a proper assessment of the axial residual stresses at the surface, three measurements along the circumference in a distance of 120° are performed. The measurements are conducted before testing and clamping. For further analysis, the mean value is considered to serve as a base value with the scatter band representing a confidence level of 95%. The residual stress results are normalized to the UTS of the material and abbreviated as $\sigma_{res,ax,surf}$. This allows for quantifying the intensity of residual stresses as a share of the ultimate tensile strength and enables a sophisticated valuation of the range in which the occurring stresses lie. All measured stresses are in the tensile region. The analysis reveals a significant decrease of residual stresses for both post treated conditions referring to the AB condition. It is found that HIPing reduces the axial residual stresses at the surface by 54.2% and solution annealing by 46.7%. Each specimen which reached the run-out criterion was measured again and showed no change. The outcome of the in situ residual stress measurements validate that testing at the fatigue limit (run-out load level) causes no notable changes of surface residual stresses. This case is depicted by the two black lines in Figure 6. However, increasing the tensile load above the fatigue limit either leads to a relaxation of residual stresses or failure before measurable changes to the residual stress state; see red lines in Figure 6, occur. The findings therefore prove that residual stresses measured before testing are still present after testing at run-out level or remain even unchanged until failure. This enables to look at measured values before testing as permanent present mean stresses. All results are given in Table 3, whereby all stress values are normalized to the surface stress before testing but after the specimen is clamped.

Table 3. Axial surface residual stress measurement results.

Condition	Surface	$\sigma_{res,ax,surf}$ (0 LC)	$\sigma_{res,ax,surf}$ (1E7 LC)	Difference
AB	UP	0.107 ± 0.027	0.106 ± 0.023	-0.9%
HIP	UP	0.049 ± 0.023	0.054 ± 0.024	$+11.0\%$
SA	UP	0.057 ± 0.026	-	

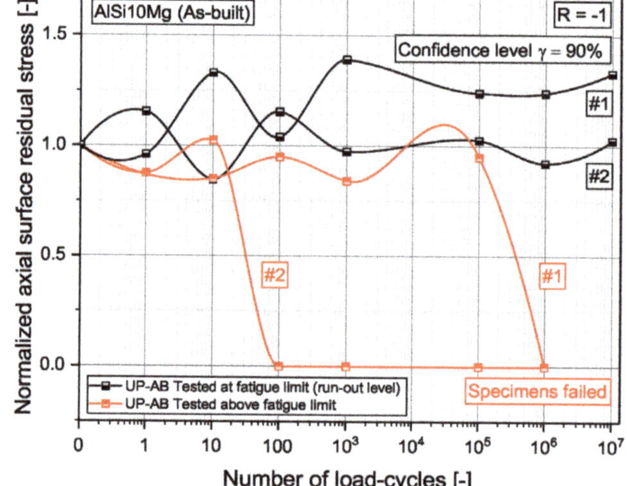

Figure 6. Cyclic stability of residual stresses.

3.2.2. In-Depth Residual Stress Distribution

To characterize the residual stress state directly at the crack initiation site for the machined specimens, it is necessary to electrolytically polish into the depth in which the failure responsible defects lie. The determination of the residual stresses at the crack origin is essential since they are substantially involved in failure initiation and crack growth; the present stress is denoted in the following as $\sigma_{res,ax,surf}$ for crack initiation at the surface and $\sigma_{res,ax,bulk}$ for failure from internal defects. To negate the effect of machining, an in-depth progression of residual stresses of the AB and HIP condition is performed. Based on the fracture surface analysis of the machined specimens, it is found that the average failure critical imperfection either lies at the surface or in a maximum depth of about 200 µm beneath the surface. Considering this, a conservative assumption is made to take the mean residual stress estimated within the aforementioned region for further analysis. The in-depth progression is shown in Figure 7, in which all stress values are normalized to the respective stress measured at the surface to highlight the distribution of residual stresses in depth. The greyed out area marks the machining allowance of 1 mm that is added to the building process. Beneath the unprocessed surface, a stress peak is observed for the HIP and AB conditions. Both show a similar progression with significantly increased axial tensile stresses in the area in which the critical imperfections lie, signalized by the red-shaded area. The results are summarized in Table 4.

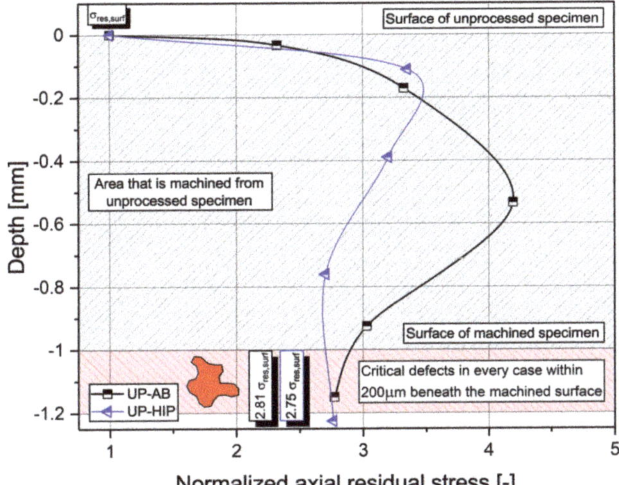

Figure 7. Normalized in-depth axial residual stress progression of AB and HIP conditions.

Table 4. Axial in-depth residual stress measurement results.

Condition	$\sigma_{res,ax,bulk}$ to $\sigma_{res,ax,surf}$	$\sigma_{res,ax,surf}$	$\sigma_{res,ax,bulk}$	Increase
UP-AB	2.81	0.107	0.301	+281%
UP-HIP	2.75	0.049	0.135	+275%

Considering the comparably high residual stresses at the crack initiation spot as an existing mean stress, they change the present mean stress state and affect the crack initiation, propagation and consequently the fatigue strength in great measure [69,70].

3.3. Surface Roughness Parameter Evaluation

For the application of the notch effect concept by Peterson, mean values of all gathered data of S_t and ρ are taken into the calculation of K_t, since the most critical surface feature is a certain combination of notch depth and notch valley radius. Since the aim is to non-destructively determine the reduction in fatigue strength, the values for S_t and ρ are taken from the optical surface assessment and not from a subsequently performed fracture surface analysis. Empirical investigations show that the mean value of the maximum valley depth of all 12 segments describes the critical surface roughness properly. For a suitable assessment of the area-based roughness parameter S_t, comparison, and validation of the optical evaluation, the maximum surface deviation is also measured within the fractured surfaces. The non-destructive optical surface evaluation is in sound correlation with the mean values from measurements on fractured specimens. The average deviation of the two methods varies between 5.8% and 7.4%, which confirms the applicability of the used evaluation routine. The results for the surface roughness parameter S_t are normalized to the mean value evaluated by the fracture surface analysis and are summarized in Table 5. It is observed that both post treatments have a beneficial impact on the surface roughness; S_t is decreased by about 14%.

The specimens are printed in a vertical (axial) direction, which leads to a periodically repetitive formation of the surface shape in the building direction. This recurring surface texture for additively manufactured structures is also reported in [71]. Three-dimensional surface imaging allows the measurement of the recurring roughness valley radii (ρ) in the loading direction with only minor deviations; see Figure 8. The evaluation is based on line measurements at several selected specimens and different locations around each specimen. It is mentioned that the notch radii can not be measured in the fractured surface since this would provide the notch radius within the wrong plane, namely perpendicular to the loading direction. The comparison of the investigated conditions reveals that the average roughness valley radius increases due to the post treatments, which mitigates the sharpness of the notch.

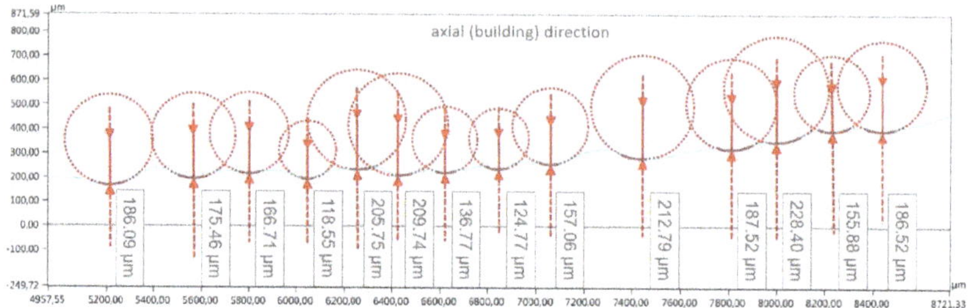

Figure 8. Surface notch valley radii measurement.

Table 5. Results of the surface roughness evaluation.

Condition	Norm. Mean S_t (Frac. Surf.)	Norm. Mean S_t (Optical Eval.)	Deviation	Average ρ
AB	1.000 (Basis)	0.926 (−7.4%)	7.4%	197.6 µm
HIP	0.868 (−13.2%)	0.804 (−19.4%)	6.2%	243.2 µm
SA	0.852 (−14.8%)	0.794 (−20.6%)	5.8%	245.5 µm

3.4. High Cycle Fatigue Testing

The high-cycle fatigue test results for the HIP condition are displayed in Figure 9. The solid lines denote the machined surface condition, whereby black with square markings represents the AB condition and blue with triangle markers is used for the HIP condition. Solely, the comparison of both machined HIP to AB conditions is published within a previous study in [35]. The dashed lines stand for the unprocessed surface condition. The displayed SN-curves are evaluated at a survival probability of 50%. All results are summarized in Table 6. The finite life region is denoted as FLR, and the long life region is abbreviated as LLR. In order to obtain reasonable results and ensure testing within the linear-elastic region, the peak load level for testing is below the yield strength of the material.

Comparing the machined conditions, the HIP treatment leads to an increase in fatigue strength by 13.8% referring to the AB condition. A similar trend is observed for the unprocessed condition. The HIPed series exhibits a 25.3% higher fatigue strength than the AB series. For both post treatment conditions, the difference between machined and unprocessed surface condition is significant. The as-built surface decreases the fatigue strength for the HIP condition by 62.2% and by 65.6% for the AB condition. Hence, the assessment of the surface roughness is essential. Regarding the scattering between 10% and 90% survival probability, HIPing narrows the scatter band for each surface condition within the finite life region as well as in the long life region. It is observed that the HIP treatment also positively impacts the slope of the S/N-curves in terms of a less steep behaviour. Partially, these results are already published in [35].

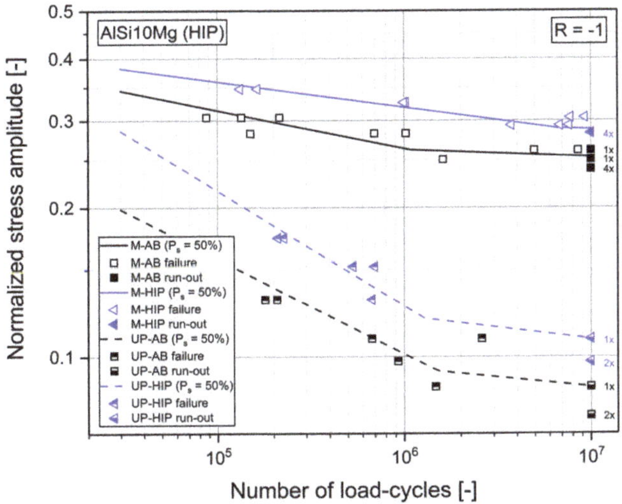

Figure 9. S/N curves for the AB and HIP test series.

The following Figure 10 shows the fatigue test results for the solution annealed condition. As described before, black lines and markings refer to the AB condition. Analogous to Figure 9, the green solid line presents the results for the machined, and the green dashed line the results of the unprocessed condition. Green circular markings are used to flag the test data. Solution annealing reveals the same trend as observed for the HIP condition. The fatigue strength of the machined SA condition lies 5.9% above the fatigue strength of the machined AB. In regard to the unprocessed surface condition, solution annealing enhances the fatigue strength by 25.3%. One can observe that the unprocessed surface again has a major impact on the fatigue behaviour, as machining leads to an improvement of +146%. The scattering between 10% and 90% survival probability is again decreased for the machined condition. The slope in the finite life region is again found to be less steep than for the AB condition.

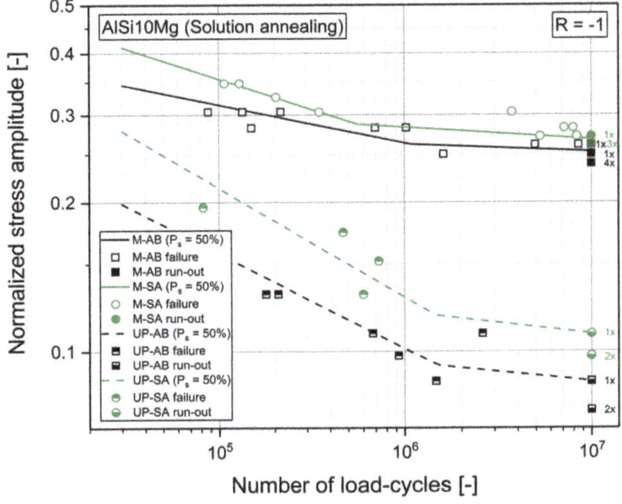

Figure 10. S/N curves for the AB and SA test series.

Table 6. High cycle fatigue test results.

Condition	Surface	$\sigma_{f(50\%)}$	Comparing AB-M	Comparing M and UP
AB	M	0.253	Basis	Basis
AB	UP	0.087	−65.6%	−65.6%
HIP	M	0.288	+13.8%	Basis
HIP	UP	0.109	−56.9%	−62.2%
SA	M	0.268	+5.9%	Basis
SA	UP	0.109	−56.9%	−59.3%

Condition	Surface	Slope FLR	Scatter Band FLR	Scatter Band LLR
AB	M	12.99	1:1.15	1:1.14
AB	UP	5.20	1:1.44	1:1.57
HIP	M	19.37	1:1.06	1:1.04
HIP	UP	4.30	1:1.22	1:1.43
SA	M	8.17	1:1.03	1:1.07
SA	UP	4.54	1:1.53	1:1.43

3.5. Fracture Surface Analysis

In order to holistically characterize the fatigue behaviour of the investigated material, a fracture surface analysis is carried out for every tested specimen. It is found that there are different mechanisms that cause the failure.

3.5.1. Failure from Intrinsic Imperfections

Investigating the fractured surfaces of the machined AB condition reveals that, in every case, surface-near pores are responsible for failure; see Figure 11a. The size and location of the imperfection are the determining criteria in terms of the fatigue strength [72–74]. For the machined HIP test series, the failure initiates from microstructural inhomogeneities. The debonding of Si-crystals is responsible for crack initiation, which is depicted in Figure 11b. This failure behaviour is already published within preliminary studies on this topic [35]. The post treatment of the SA condition is similar to the HIP treatment, which leads to a comparable microstructure. On the contrary, the fracture surface analysis displays a combined failure cause of microstructural inhomogeneities and porosity, as shown in Figure 11d. The occurring porosity may be attributed to the lack of isostatic pressure during the SA treatment. To be sure about the failure mechanism, an EDX-Analysis is performed on the fractured surface. In regard to Figure 11c, area 'a' shows a chemical composition of $Al_{18.06}Si_{65.41}Mg_{16.53}$. Spots 'b' and 'c' consist of a great measure of Silicon, which leads to the interpretation of debonding Si-crystals, also found in [66]. In comparison, spot 'd', which lies beneath a delaminated Si-Slab, is found to be base material.

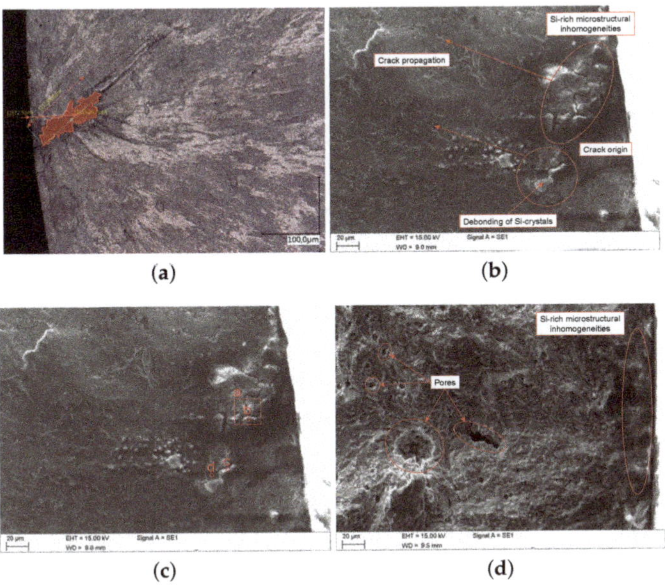

Figure 11. Fracture surface analysis of machined specimens. (**a**) Failure initiation spot of AB specimens. (**b**) Failure initiation spot of HIP specimens. (**c**) EDX analysis on the fractured surface of one HIPed specimen. (**d**) Failure initiation spot of SA specimens.

3.5.2. Failure from Surface Features

The main outcome of the fracture surface analysis for all test series and each specimen exhibiting an unprocessed surface is that the surface texture is in every case failure critical. The effect of the surface roughness dominates all other imperfections and microstructural features in terms of crack initiation and the consequential fatigue strength. This behaviour is also observed in [75]. Figure 12a,b highlight the failure origin from a roughness valley. The substantive effect of the surface roughness on the fatigue strength is well reported in [76–78]. The given examples are from the unprocessed AB series. No evidence of pores or microstructural inhomogeneities is found in the surrounding area for any test series. In conclusion, one can distinctively determine the surface condition as the crucial feature, which overshadows all other failure reasons and are therefore neglectable in the presence of an unprocessed surface.

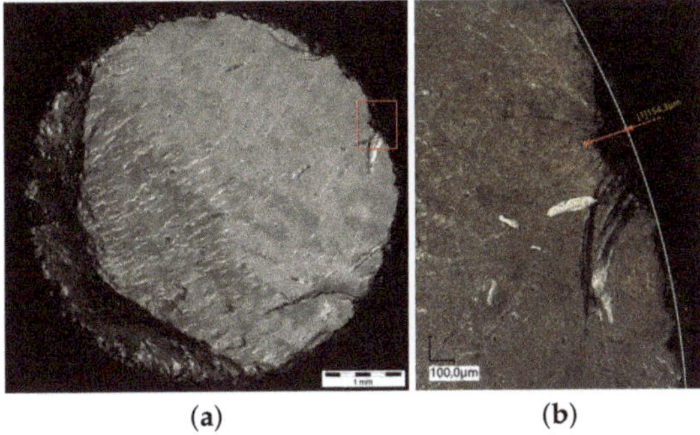

Figure 12. Fracture surface analysis for one specimen of the unprocessed condition. (**a**) Fractured surface of unprocessed as-built specimen. (**b**) Failure responsible surface characteristic.

3.6. Fatigue Assessment

3.6.1. Mean Stress Correction

Macroscopic residual stresses of the first order may be considered to overlay with load stresses and therefore act as mean stresses, encouraging a shift of the actual load stress ratio to an effective stress ratio R_{eff} [79,80]. The intended testing is performed at a load stress ratio of $R = -1$, which means that the mean stress is zero. Taking the effective mean stress caused by load and residual stresses into account, the load stress R-ratio is shifted to an effective R-ratio, according to Equation (4):

$$R_{eff} = \frac{\sigma_{min} + \sigma_{res,ax}}{\sigma_{max} + \sigma_{res,ax}}.$$ (4)

For the HIP condition, the present residual stresses lead to an effective stress ratio of $R_{eff} = -0.36$ for the machined and to $R_{eff} = -0.38$ for the unprocessed surface condition. The effective stress ratio for the AB machined condition calculates to $R_{eff} = 0.09$ and even to $R_{eff} = 0.1$ with an unprocessed surface. Hence, it is clearly shown that residual stresses alter the testing condition significantly. To independently assess the impact of the surface roughness, the stress amplitude is extrapolated to a ratio of $R = -1$. The aim is to eliminate all influencing factors but one, the surface roughness. This enables the independent quantification of it. This correction of the stress amplitude to a mean stress of zero accounts for the influence of residual stresses and simultaneously gives a conservative estimation of the endurable fatigue strength amplitude as if no residual stresses would be present. Figure 13 presents the mean stress corrected fatigue strength amplitude according to Gerber, which is denoted as $\sigma_{f,M,cor,G}$ in the following. The same procedure is applied for the correction according to Dietmann, denoted as $\sigma_{f,M,cor,D}$, shown in Figure 14. The results are also summarized in Table 7. Comparing both concepts, the model according to Gerber is more conservative than the Dietmann one in regard to the experimental results $\sigma_{f,exp}$. In conclusion, one can state that it is proven that the residual stress state contributes in great measure to the fatigue resistance; this effect can be observed by the increase of the endurable fatigue strength amplitude for the AB and HIP condition.

The difference in the residual stress free state between AB und HIP may be attributed to beneficial microstructural changes and the different failure initiation modes for the HIP condition, as previously presented and published within [35]. Both concepts lead to similar results, estimating a benefit due to HIPing of approximately +5.8% for the machined and 23.9% for the unprocessed condition, see Table 8.

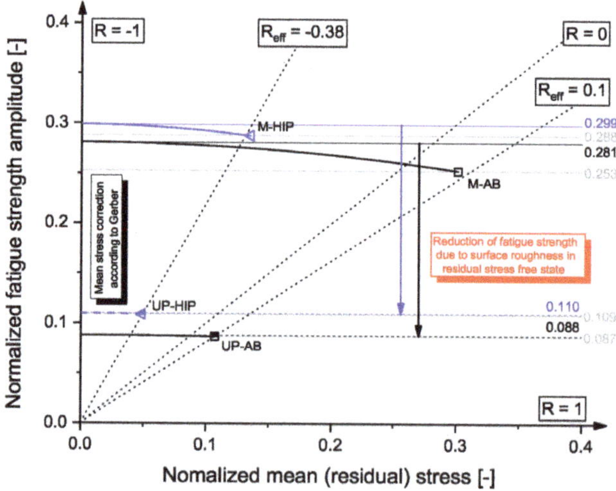

Figure 13. Haigh diagram with residual stresses accounted for according to Gerber.

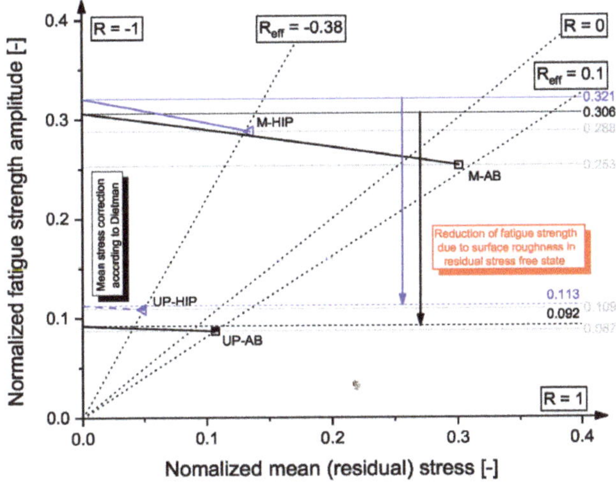

Figure 14. Haigh diagram with residual stresses accounted for according to Dietmann.

Table 7. Mean stress corrected fatigue strength values.

Condition	R_{eff}	$\sigma_{f,exp}$	$\sigma_{f,cor,G}$	$\sigma_{f,cor,D}$
M-HIP	−0.36	0.288	0.299 (+3.8%)	0.321 (+11.4%)
UP-HIP	−0.38	0.109	0.110 (+0.9%)	0.113 (+3.7%)
M-AB	0.09	0.253	0.281 (+11.1%)	0.306 (+20.9%)
UP-AB	0.10	0.087	0.088 (+1.0%)	0.092 (+5.7%)

Table 8. Impact of the microstructure on the fatigue strength in residual stress free state.

Condition	M-HIP to M-AB	UP-HIP to UP-AB
$\sigma_{f,cor,G}$	1.064 (+6.4%)	1.250 (+25.0%)
$\sigma_{f,cor,D}$	1.052 (+5.2%)	1.228 (+22.8%)

3.6.2. Assessment of the Surface Roughness in Mean Stress Corrected State

The importance of the assessment of the surface roughness caused by the building process is obvious, since it is unequivocally found to be the fatigue strength determining factor. The fatigue test results as well as the fracture surface analysis emphasize the evaluation of the surface roughness and its influence. The results for the notch factor of all conditions are given in Tables 9 and 10, in which the estimated fatigue strength based on the analytical model is abbreviated as $\sigma_{f,UP,mod}$, and the experimental results are denoted as $\sigma_{f,UP,exp}$, respectively, for each unprocessed condition. As expected based on the roughness parameters, the notch effect is more pronounced for the AB condition than for the post treated conditions. Beginning with the corrected fatigue strength of the machined condition ($\sigma_{f,M,cor}$) and dividing it by the notch factor (K_t), which acts as a reduction factor accounting for the surface roughness, estimates the fatigue strength of the unprocessed condition; see Equation (5):

$$\sigma_{f,UP,mod} = \frac{\sigma_{f,M,cor}}{K_t}.$$ (5)

Eventually, the analytically estimated, mean stress corrected fatigue strength is compared to the experimentally determined fatigue strength, both in a residual stress freed state. The results of the analytical approach deviate in the range of +6.4% to +16.3% from the experimental results, which acknowledges the applied procedure to be deployable for the estimation of the reduction of fatigue properties due to the surface roughness starting from a machined surface condition in a residual stress freed state utilizing mean stress corrected values according to Gerber, see Table 9 and Dietmann, summarized in Table 10.

Table 9. Assessment of the surface roughness on the fatigue strength after Gerber.

Condition	$\sigma_{f,M,exp}$	$\sigma_{f,M,cor,G}$	K_t (UP)	$\sigma_{f,UP,mod}$
AB	0.253	0.281	2.86	0.098
HIP	0.288	0.299	2.56	0.117

Condition	$\sigma_{f,UP,exp}$	$\sigma_{f,UP,cor,G}$	$\sigma_{f,UP,mod}$ to $\sigma_{f,UP,cor,G}$	Difference
AB	0.087	0.088	1.114	+11.4%
HIP	0.109	0.110	1.064	+6.4%

Table 10. Assessment of the surface roughness on the fatigue strength after Dietmann.

Condition	$\sigma_{f,M,exp}$	$\sigma_{f,M,cor,D}$	K_t (UP)	$\sigma_{f,UP,mod}$
AB	0.253	0.306	2.86	0.107
HIP	0.288	0.321	2.56	0.125

Condition	$\sigma_{f,UP,exp}$	$\sigma_{f,UP,cor,D}$	$\sigma_{f,UP,mod}$ to $\sigma_{f,UP,cor,D}$	Difference
AB	0.087	0.092	1.163	+16.3%
HIP	0.109	0.113	1.106	+10.6%

Both concepts present a minor non-conservative approach, but the scatter band (1:Ts) in the long life region of 1:57 for the UP-AB, and 1:43 for the UP-HIP condition, as given in Table 6, needs to be considered as well. Consequently, the estimated mean fatigue strength is well within the scattering of the experimental results.

The above presented concept is utilized to predict the fatigue strength of the SA condition. Both of the others, AB and HIP, reveal in machined and unprocessed conditions the same effective stress ratio due to residual stresses because only the residual stresses in unprocessed SA conditions are measured, assuming the same R-ratio in machined conditions. Applying this procedure, the fatigue strength of the machined SA condition can be properly predicted with both concepts, denoted as $\sigma_{f,M,pred,G/D}$. The deviation from the experimental results is calculated to only +3.4%; see Table 11.

Table 11. Fatigue strength assessment of the SA condition.

Condition	$R_{eff,M,UP}$	$\sigma_{f,UP,exp}$	$\sigma_{f,UP,cor,G}$	$\sigma_{f,UP,cor,D}$	K_t (UP)
SA	−0.31	0.109	0.110	0.114	2.54

$\sigma_{f,M,cor,G}$	$\sigma_{f,M,cor,D}$	$\sigma_{f,M,pred,G}$	$\sigma_{f,M,pred,D}$	$\sigma_{f,M,exp}$
0.279	0.290	0.277 (+3.4%)	0.277 (+3.4%)	0.268 (Base)

4. Discussion

Based on the results presented in this paper, the fatigue strength of additively manufactured AlSi10Mg structures is altered by post treatments, the residual stress state and the surface condition. The fatigue strength is improved by HIPing and solution annealing, for a machined as well as a unprocessed surface, compared to the AB condition. This study also proves a beneficial effect of the investigated post treatments on the microstructure and consequently on fatigue.

The outcome of the investigations on the surface condition reveals that, by virtue of the roughness, fatigue properties are significantly reduced. Comparing the as-built surface to a machined surface, this work reveals that the unprocessed surface causes a significant reduction of fatigue properties of about −60%. The surface roughness analysis shows that the HIP as well as the SA treatment positively influences decisive surface related characteristics due to the heat input and the applied pressure during the HIP process. The maximum roughness valley depth is decreased and furthermore the average roughness valley radius is mitigated compared to the AB condition. These beneficial changes to the surface topography contribute to an improved fatigue behaviour of +25.3% for both conditions compared to the AB condition.

This work leads to the conclusion that the residual stress state at the respective failure origin can be considered as a present mean stress, whereby a shift of the intended load stress ratio to an effective stress ratio occurs. Another finding of the conducted investigations is that, due to the heat influence of the post treatments, residual stresses are reduced by roughly 50%. An analysis of the in-depth progression reveals increased tensile residual stresses compared to the surface by a factor of almost three. By the means of the presented methodology, a prediction of the reduced fatigue strength of unprocessed specimen, in relation to the machined condition, is given. The developed model is shown to be well applicable to the investigated test series in a residual stress free state. Although the fatigue strength amplitude prediction is slightly non-conservative, the estimation is well within the scatter band of the the experimental results in the long life region.

Author Contributions: Conceptualization, W.S. and M.L.; methodology, W.S. and M.L.; validation, W.S. and M.L.; formal analysis, W.S.; investigation, W.S., S.P. and S.S.; resources, W.S.; data curation, W.S. and F.B.; writing—original draft preparation, W.S.; writing—review and editing, W.S. and M.L.; visualization, W.S.; supervision, M.L. and F.G.; project administration, M.L. and F.G.

Funding: This research received no external funding.

Acknowledgments: Special thanks are givento the Austrian Research Promotion Agency (FFG), who funded the research project by funds of the Federal Ministry for Transport, Innovation and Technology (bmvit) and the Federal Ministry for Digital and Economic Affairs (bmdw). Scientific support regarding the optical surface topography measurement and evaluation was provided in the course of the "Christian Doppler Laboratory for Manufacturing Process based Component Design".

Conflicts of Interest: The authors declare no conflict of interest.

References

1. Leary, M.; Mazur, M.; Elambasseril, J.; McMillan, M.; Chirent, T.; Sun, Y.; Qian, M.; Easton, M.; Brandt, M. Selective laser melting (SLM) of AlSi12Mg lattice structures. *Mater. Des.* **2016**, *98*, 344–357. [CrossRef]

2. Brandt, M.; Sun, S.J.; Leary, M.; Feih, S.; Elambasseril, J.; Liu, Q.C. High-Value SLM Aerospace Components: From Design to Manufacture. *Adv. Mater. Res.* **2013**, *633*, 135–147. [CrossRef]

3. Wang, X.; Xu, S.; Zhou, S.; Xu, W.; Leary, M.; Choong, P.; Qian, M.; Brandt, M.; Xie, Y.M. Topological design and additive manufacturing of porous metals for bone scaffolds and orthopaedic implants: A review. *Biomaterials* **2016**, *83*, 127–141. [CrossRef] [PubMed]

4. Harun, W.; Kamariah, M.; Muhamad, N.; Ghani, S.; Ahmad, F.; Mohamed, Z. A review of powder additive manufacturing processes for metallic biomaterials. *Powder Technol.* **2018**, *327*, 128–151. [CrossRef]

5. Huynh, L.; Rotella, J.; Sangid, M.D. Fatigue behavior of IN718 microtrusses produced via additive manufacturing. *Mater. Des.* **2016**, *105*, 278–289. [CrossRef]

6. Koutiri, I.; Pessard, E.; Peyre, P.; Amlou, O.; de Terris, T. Influence of SLM process parameters on the surface finish, porosity rate and fatigue behavior of as-built Inconel 625 parts. *J. Mater. Process. Technol.* **2018**, *255*, 536–546. [CrossRef]

7. Vayssette, B.; Saintier, N.; Brugger, C.; Elmay, M.; Pessard, E. Surface roughness of Ti-6Al-4V parts obtained by SLM and EBM: Effect on the High Cycle Fatigue life. *Procedia Eng.* **2018**, *213*, 89–97. [CrossRef]

8. Campbell, G.; Lahey, R. A survey of serious aircraft accidents involving fatigue fracture. *Int. J. Fatigue* **1984**, *6*, 25–30. [CrossRef]

9. Uzan, N.E.; Shneck, R.; Yeheskel, O.; Frage, N. Fatigue of AlSi10Mg specimens fabricated by additive manufacturing selective laser melting (AM-SLM). *Mater. Sci. Eng. A* **2017**, *704*, 229–237. [CrossRef]

10. Aboulkhair, N.T.; Maskery, I.; Tuck, C.; Ashcroft, I.; Everitt, N.M. Improving the fatigue behaviour of a selectively laser melted aluminium alloy: Influence of heat treatment and surface quality. *Mater. Des.* **2016**, *104*, 174–182. [CrossRef]

11. Afkhami, S.; Dabiri, M.; Alavi, S.H.; Björk, T.; Salminen, A. Fatigue characteristics of steels manufactured by selective laser melting. *Int. J. Fatigue* **2019**, *122*, 72–83. [CrossRef]

12. Liu, Y.J.; Li, S.J.; Wang, H.L.; Hou, W.T.; Hao, Y.L.; Yang, R.; Sercombe, T.B.; Zhang, L.C. Microstructure, defects and mechanical behavior of beta-type titanium porous structures manufactured by electron beam melting and selective laser melting. *Acta Mater.* **2016**, *113*, 56–67. [CrossRef]

13. Dai, D.; Gu, D. Effect of metal vaporization behavior on keyhole-mode surface morphology of selective laser melted composites using different protective atmospheres. *Appl. Surf. Sci.* **2015**, *355*, 310–319. [CrossRef]

14. Gaytan, S.M.; Murr, L.E.; Medina, F.; Martinez, E.; Lopez, M.I.; Wicker, R.B. Advanced metal powder based manufacturing of complex components by electron beam melting. *Mater. Technol.* **2013**, *24*, 180–190. [CrossRef]

15. Zhang, L.C.; Attar, H. Selective Laser Melting of Titanium Alloys and Titanium Matrix Composites for Biomedical Applications: A Review. *Adv. Eng. Mater.* **2016**, *18*, 463–475. [CrossRef]

16. Zhao, S.; Li, S.J.; Wang, S.G.; Hou, W.T.; Li, Y.; Zhang, L.C.; Hao, Y.L.; Yang, R.; Misra, R.; Murr, L.E. Compressive and fatigue behavior of functionally graded Ti-6Al-4V meshes fabricated by electron beam melting. *Acta Mater.* **2018**, *150*, 1–15. [CrossRef]

17. Liu, Y.J.; Wang, H.L.; Li, S.J.; Wang, S.G.; Wang, W.J.; Hou, W.T.; Hao, Y.L.; Yang, R.; Zhang, L.C. Compressive and fatigue behavior of beta-type titanium porous structures fabricated by electron beam melting. *Acta Mater.* **2017**, *126*, 58–66. [CrossRef]

18. Ceschini, L.; Morri, A.; Sambogna, G. The effect of hot isostatic pressing on the fatigue behaviour of sand-cast A356-T6 and A204-T6 aluminum alloys. *J. Mater. Process. Technol.* **2008**, *204*, 231–238. [CrossRef]

19. Domfang Ngnekou, J.N.; Nadot, Y.; Henaff, G.; Nicolai, J.; Ridosz, L. Influence of defect size on the fatigue resistance of AlSi10Mg alloy elaborated by selective laser melting (SLM). *Procedia Struct. Integr.* **2017**, *7*, 75–83. [CrossRef]

20. Domfang Ngnekou, J.N.; Nadot, Y.; Henaff, G.; Nicolai, J.; Kan, W.H.; Cairney, J.M.; Ridosz, L. Fatigue properties of AlSi10Mg produced by Additive Layer Manufacturing. *Int. J. Fatigue* **2019**, *119*, 160–172. [CrossRef]

21. Lee, M.H.; Kim, J.J.; Kim, K.H.; Kim, N.J.; Lee, S.; Lee, E.W. Effects of HIPping on high-cycle fatigue properties of investment cast A356 aluminum alloys. *Mater. Sci. Eng. A* **2003**, *340*, 123–129. [CrossRef]

22. Wang, Q.; Apelian, D.; Lados, D. Fatigue behavior of A356-T6 aluminum cast alloys. Part I. Effect of casting defects. *J. Light Met.* **2001**, *1*, 73–84. [CrossRef]

23. Wang, Q.; Apelian, D.; Lados, D. Fatigue behavior of A356/357 aluminum cast alloys. Part II – Effect of microstructural constituents. *J. Light Met.* **2001**, *1*, 85–97. [CrossRef]

24. Herzog, D.; Seyda, V.; Wycisk, E.; Emmelmann, C. Additive manufacturing of metals. *Acta Mater.* **2016**, *117*, 371–392. [CrossRef]

25. Beretta, S.; Romano, S. A comparison of fatigue strength sensitivity to defects for materials manufactured by AM or traditional processes. *Int. J. Fatigue* **2017**, *94*, 178–191. [CrossRef]

26. Buffiere, J.Y. Fatigue Crack Initiation In addition, Propagation From Defects In Metals: Is 3D Characterization Important? *Procedia Struct. Integr.* **2017**, *7*, 27–32. [CrossRef]

27. Ngnekou, J.N.D.; Henaff, G.; Nadot, Y.; Nicolai, J.; Ridosz, L. Fatigue resistance of selectively laser melted aluminum alloy under T6 heat treatment. *Procedia Eng.* **2018**, *213*, 79–88. [CrossRef]

28. Brandl, E.; Heckenberger, U.; Holzinger, V.; Buchbinder, D. Additive manufactured AlSi10Mg samples using Selective Laser Melting (SLM): Microstructure, high cycle fatigue, and fracture behavior. *Mater. Des.* **2012**, *34*, 159–169. [CrossRef]

29. Arola, D.; Williams, C. Estimating the fatigue stress concentration factor of machined surfaces. *Int. J. Fatigue* **2002**, *24*, 923–930. [CrossRef]

30. Peterson, R.E.; Plunkett, R. Stress Concentration Factors. *J. Appl. Mech.* **1975**, *42*, 248. [CrossRef]

31. Webster, G.A.; Ezeilo, A.N. Residual stress distributions and their influence on fatigue lifetimes. *Int. J. Fatigue* **2001**, *23*, 375–383. [CrossRef]

32. Zhuang, W.Z.; Halford, G.R. Investigation of residual stress relaxation under cyclic load. *Int. J. Fatigue* **2001**, *23*, 31–37. [CrossRef]

33. EOS GmbH—Electro Optical Systems. *Data Sheet: EOS Aluminium AlSi10Mg*; EOS GmbH—Electro Optical Systems: Krailling, Germany, 2014.

34. European Comittee for Standardization (CEN). *Aluminium and Aluminium Alloys—Castings—Chemical Composition and Mechanical Properties*; European Comittee for Standardization: Brussels, Belgium, 2010.

35. Schneller, W.; Leitner, M.; Springer, S.; Grün, F.; Taschauer, M. Effect of HIP Treatment on Microstructure and Fatigue Strength of Selectively Laser Melted AlSi10Mg. *J. Manuf. Mater. Process.* **2019**, *3*, 16. [CrossRef]

36. ASTM. *ASTM E1508-98 (2003), Standard Guide for Quantitative Analysis by Energy-Dispersive Spectroscopy*; ASTM: West Conshohocken, PA, USA, 2012.

37. Takata, N.; Kodaira, H.; Sekizawa, K.; Suzuki, A.; Kobashi, M. Change in microstructure of selectively laser melted AlSi10Mg alloy with heat treatments. *Mater. Sci. Eng. A* **2017**, *704*, 218–228. [CrossRef]

38. Li, W.; Li, S.; Liu, J.; Zhang, A.; Zhou, Y.; Wei, Q.; Yan, C.; Shi, Y. Effect of heat treatment on AlSi10Mg alloy fabricated by selective laser melting: Microstructure evolution, mechanical properties and fracture mechanism. *Mater. Sci. Eng. A* **2016**, *663*, 116–125. [CrossRef]

39. Tensi, H.M.; Hogerl, J. Influence of Heat Treatment on the Microstructure and Mechanical Behavior of High Strength AlSi Cast Alloys. In *Heat Treating, Proceedings of the 16th Conference*; Dossett, J.L., Luetje, R.E., Eds.; ASM International: Materials Park, OH, USA, 2010; pp. 243–247.

40. Cai, C.; Geng, H.; Wang, S.; Gong, B.; Zhang, Z. Microstructure Evolution of AlSi10Mg(Cu) Alloy Related to Isothermal Exposure. *Materials* **2018**, *11*, 809. [CrossRef]

41. Nicholas, T. *High Cycle Fatigue: A Mechanics of Materials Perspective*, 1st ed.; Elsevier: Amsterdam, The Netherlands, 2006.

42. Gänser, H.P.; Maierhofer, J.; Christiner, T. Statistical correction for reinserted runouts in fatigue testing. *Int. J. Fatigue* **2015**, *80*, 76–80. [CrossRef]

43. Dengel, D. Die arc sin $\sqrt{P}$-Transformation—Ein einfaches Verfahren zur grafischen und rechnerischen Auswertung geplanter Wöhlerversuche. *Mater. Werkst.* **1975**, *6*, 253–261. [CrossRef]

44. ASTM International. *Standard Practice for Statistical Analysis of Linear or Linearized Stress-Life (S-N) and Strain-Life (e-N) Fatigue Data*; ASTM: West Conshohocken, PA, USA, 2015.

45. Bader, Q. Kadum, A. Mean Stress Correction Effects On the Fatigue Life Behavior of Steel Alloys by Using Stress Life Approach Theories. *Int. J. Eng. Technol.* **2014**, *10*, 50–58.

46. Dowling, N.E.; Calhoun, C.A.; Arcari, A. Mean stress effects in stress-life fatigue and the Walker equation. *Fatigue Fract. Eng. Mater. Struct.* **2009**, *32*, 163–179. [CrossRef]

47. Pallarés-Santasmartas, L.; Albizuri, J.; Avilés, A.; Avilés, R. Mean Stress Effect on the Axial Fatigue Strength of DIN 34CrNiMo6 Quenched and Tempered Steel. *Metals* **2018**, *8*, 213. [CrossRef]

48. Gerber, H. Bestimmung der zulässigen Spannungen in Eisenkonstruktionen. *Z. Bayrischen Arch. Ing. Vereins* **1874**, *6*, 101–110.

49. Dietman, H. Festigkeitsberechnung bei Mehrachsiger Schwingbeanspruchung. *Konstruktion* **1973**, *25*, 181–189.

50. Mercelis, P.; Kruth, J.P. Residual stresses in selective laser sintering and selective laser melting. *Rapid Prototyp. J.* **2006**, *12*, 254–265. [CrossRef]

51. Wang, L.; Jiang, X.; Zhu, Y.; Ding, Z.; Zhu, X.; Sun, J.; Yan, B. Investigation of Performance and Residual Stress Generation of AlSi10Mg Processed by Selective Laser Melting. *Adv. Mater. Sci. Eng.* **2018**, *2018*, 1–12. [CrossRef]

52. ASTM. *ASTM, E915-96(2002), Standard Test Method for Verifying the Alignment of X-ray Diffraction Instrumentation for Residual Stress Measurement*; ASTM International: West Conshohocken, PA, USA, 2012.

53. ASTM. *ASTM, E2860-12, Standard Test Method for Residual Stress Measurement by X-ray Diffraction for Bearing Steels*; ASTM International: West Conshohocken, PA, USA, 2012.

54. DIN. *DIN EN ISO 4287: 2010-07, Geometrische Produktspezifikation (GPS)—Oberflächenbeschaffenheit: Tastschnittverfahren— Benennungen, Definitionen und Kenngrößen der Oberflächenbeschaffenheit*; Beuth: Berlin, Germany, 2010. [CrossRef]

55. Pomberger, S.; Stoschka, M.; Leitner, M. Cast surface texture characterisation via areal roughness. *Precis. Eng.* **2019**, *60*, 465–481. [CrossRef]

56. Tang, M.; Pistorius, P.C.; Narra, S.; Beuth, J.L. Rapid Solidification: Selective Laser Melting of AlSi10Mg. *JOM* **2016**, *68*, 960–966. [CrossRef]

57. Dai, D.; Gu, D.; Zhang, H.; Zhang, J.; Du, Y.; Zhao, T.; Hong, C.; Gasser, A.; Poprawe, R. Heat-induced molten pool boundary softening behavior and its effect on tensile properties of laser additive manufactured aluminum alloy. *Vacuum* **2018**, *154*, 341–350. [CrossRef]

58. Islam, M.A.; Farhat, Z.N. The influence of porosity and hot isostatic pressing treatment on wear characteristics of cast and P/M aluminum alloys. *Wear* **2011**, *271*, 1594–1601. [CrossRef]

59. Bösch, D.; Pogatscher, S.; Hummel, M.; Fragner, W.; Uggowitzer, P.J.; Göken, M.; Höppel, H.W. Secondary Al-Si-Mg High-pressure Die Casting Alloys with Enhanced Ductility. *Metall. Mater. Trans. A* **2015**, *46*, 1035–1045. [CrossRef]

60. Mulazimoglu, M.H.; Zaluska, A.; Gruzleski, J.E.; Paray, F. Electron microscope study of Al-Fe-Si intermetallics in 6201 aluminum alloy. *Metall. Mater. Trans. A* **1996**, *27*, 929–936. [CrossRef]

61. Zhou, L.; Mehta, A.; Schulz, E.; McWilliams, B.; Cho, K.; Sohn, Y. Microstructure, precipitates and hardness of selectively laser melted AlSi10Mg alloy before and after heat treatment. *Mater. Charact.* **2018**, *143*, 5–17. [CrossRef]

62. Maamoun, A.H.; Elbestawi, M.; Dosbaeva, G.K.; Veldhuis, S.C. Thermal post-processing of AlSi10Mg parts produced by Selective Laser Melting using recycled powder. *Addit. Manuf.* **2018**, *21*, 234–247. [CrossRef]

63. Prashanth, K.G.; Scudino, S.; Klauss, H.J.; Surreddi, K.B.; Löber, L.; Wang, Z.; Chaubey, A.K.; Kühn, U.; Eckert, J. Microstructure and mechanical properties of Al–12Si produced by selective laser melting: Effect of heat treatment. *Mater. Sci. Eng. A* **2014**, *590*, 153–160. [CrossRef]

64. Fousová, M.; Dvorský, D.; Michalcová, A.; Vojtěch, D. Changes in the microstructure and mechanical properties of additively manufactured AlSi10Mg alloy after exposure to elevated temperatures. *Mater. Charact.* **2018**, *137*, 119–126. [CrossRef]

65. Chrominski, W.; Lewandowska, M. Precipitation phenomena in ultrafine grained Al–Mg–Si alloy with heterogeneous microstructure. *Acta Mater.* **2016**, *103*, 547–557. [CrossRef]

66. Gall, K.; Yang, N.; Horstemeyer, M.; McDowell, D.L.; Fan, J. The debonding and fracture of Si particles during the fatigue of a cast Al-Si alloy. *Metall. Mater. Trans. A* **1999**, *30*, 3079–3088. [CrossRef]

67. Edwards, P.; Ramulu, M. Fatigue performance evaluation of selective laser melted Ti–6Al–4V. *Mater. Sci. Eng. A* **2014**, *598*, 327–337. [CrossRef]

68. Biffi, C.A.; Fiocchi, J.; Tuissi, A. Selective laser melting of AlSi10 Mg: Influence of process parameters on Mg2Si precipitation and Si spheroidization. *J. Alloys Compd.* **2018**, *755*, 100–107. [CrossRef]

69. Withers, P.J.; Bhadeshia, H. Residual stress. Part 1—Measurement techniques. *Mater. Sci. Technol.* **2013**, *17*, 355–365. [CrossRef]

70. Aigner, R.; Leitner, M.; Stoschka, M. On the mean stress sensitivity of cast aluminium considering imperfections. *Mater. Sci. Eng. A* **2019**, *758*, 172–184. [CrossRef]

71. Townsend, A.; Senin, N.; Blunt, L.; Leach, R.K.; Taylor, J.S. Surface texture metrology for metal additive manufacturing: A review. *Precis. Eng.* **2016**, *46*, 34–47. [CrossRef]

72. Masuo, H.; Tanaka, Y.; Morokoshi, S.; Yagura, H.; Uchida, T.; Yamamoto, Y.; Murakami, Y. Effects of Defects, Surface Roughness and HIP on Fatigue Strength of Ti-6Al-4V manufactured by Additive Manufacturing. *Procedia Struct. Integr.* **2017**, *7*, 19–26. [CrossRef]

73. Romano, S.; Beretta, S.; Brandão, A.; Gumpinger, J.; Ghidini, T. HCF resistance of AlSi10Mg produced by SLM in relation to the presence of defects. *Procedia Struct. Integr.* **2017**, *7*, 101–108. [CrossRef]

74. Romano, S.; Brückner-Foit, A.; Brandão, A.; Gumpinger, J.; Ghidini, T.; Beretta, S. Fatigue properties of AlSi10Mg obtained by additive manufacturing: Defect-based modelling and prediction of fatigue strength. *Eng. Fract. Mech.* **2018**, *187*, 165–189. [CrossRef]

75. Romano, S.; Brandão, A.; Gumpinger, J.; Gschweitl, M.; Beretta, S. Qualification of AM parts: Extreme value statistics applied to tomographic measurements. *Mater. Des.* **2017**, *131*, 32–48. [CrossRef]

76. Greitemeier, D.; Dalle Donne, C.; Syassen, F.; Eufinger, J.; Melz, T. Effect of surface roughness on fatigue performance of additive manufactured Ti–6Al–4V. *Mater. Sci. Technol.* **2015**, *32*, 629–634. [CrossRef]

77. Pegues, J.; Roach, M.; Scott Williamson, R.; Shamsaei, N. Surface roughness effects on the fatigue strength of additively manufactured Ti-6Al-4V. *Int. J. Fatigue* **2018**, *116*, 543–552. [CrossRef]

78. Da Silva, P.S.C.P.; Campanelli, L.C.; Escobar Claros, C.A.; Ferreira, T.; Oliveira, D.P.; Bolfarini, C. Prediction of the surface finishing roughness effect on the fatigue resistance of Ti-6Al-4V ELI for implants applications. *Int. J. Fatigue* **2017**, *103*, 258–263. [CrossRef]

79. Wolfstieg, U.; Macherauch, E. Ursachen und Bewertung von Eigenspannungen. *Chem. Ing. Tech. CIT* **1973**, *45*, 760–770. [CrossRef]

80. Kloos, K.H. Eigenspannungen, Definition und Entstehungsursachen. *Mater. Werkst.* **1979**, *10*, 293–302. [CrossRef]

MDPI

St. Alban-Anlage 66

4052 Basel

Switzerland

Tel. +41 61 683 77 34

Fax +41 61 302 89 18

www.mdpi.com

Journal of Manufacturing and Materials Processing Editorial Office

E-mail: jmmp@mdpi.com

www.mdpi.com/journal/jmmp